Pharmaceutical Quality Systems

Edited by

Oliver Schmidt

CRC Press
Taylor & Francis Group
Boca Raton London New York

CRC Press is an imprint of the
Taylor & Francis Group, an **informa** business

CRC Press
Taylor & Francis Group
6000 Broken Sound Parkway NW, Suite 300
Boca Raton, FL 33487-2742

First issued in paperback 2019

© 2010 by Taylor & Francis Group, LLC
CRC Press is an imprint of Taylor & Francis Group, an Informa business

No claim to original U.S. Government works

ISBN-13: 978-1-57491-109-1 (hbk)
ISBN-13: 978-0-367-39870-5 (pbk)

A CIP record for this book is available from the British Library.

Library of Congress Cataloging-in-Publication Data available on application

**Visit the Taylor & Francis Web site at
http://www.taylorandfrancis.com**

**and the CRC Press Web site at
http://www.crcpress.com**

Contents

Preface

Quality System—Quality Assurance System—Quality Management System: What is the correct terminology? The answer is that there is *no* correct terminology, and this is the main problem!

However, many definitions can be found within Good Manufacturing Practice (GMP) Guides of the European Commission (EC), World Health Organization (WHO) or within the ISO 9000 Standard. But they differ from each other. As a result, every pharmaceutical company has to decide to which of these definitions they will refer.

To give a helping hand to the reader of this book, chapter 1 contains a summary of the common definitions.

But the definitions are not the only problem. When a company decides to build up a Quality System, they have to face the fact that there aren't many pharmaceutical guidelines which define exactly how such system has to be built up.

We decided to ask a European GMP Inspector to interpret the existing regulatory demands on Quality Systems and how these requirements are monitored by the European Inspectorates. He also describes the different views of the Food and Drug Administration (FDA) and the EC concerning Quality Assurance functions. The EC GMP guides states "Quality Assurance is more than GMP." But what does 'more' mean? No pharmaceutical guideline answers this question. ISO 9000 reflects the worldwide standard for Quality Management and Quality Assurance Systems. It is used in many branches including the pharmaceutical industry.

Part 2 of this book will describe the contents of ISO 9000, how to apply this standard, and most importantly, how to combine GMP and ISO 9000.

Chapter 3 will explain exactly how to combine Drug GMP with ISO 9000 and chapter 4 explains how to combine Active Pharmaceutical Ingredients (API) GMP with ISO 9000. Each of these chapters includes a matrix that will show the consistencies and differences.

Many manufacturers of starting materials find it difficult to build up a company-wide GMP/ISO 9000 Quality System because only 1 percent of their total production is manufactured for customers from the pharmaceutical industry. The Master Plan Concept is a practical approach to this situation.

Part 3, Implementation Reviews, shows how Quality (Management) systems (QMS) have been installed successfully in pharmaceutical companies.

The highest documentation level in a Quality System is the so-called Quality Manual. It is an important tool for establishing a company-wide Quality System. Chapter 6 describes the model of Hoechst Marion Roussel. Chapter 7 and Chapter 8 contain the implementation reviews of two QMS which have been established successfully at Biotest AG, Germany, and Schering Produktionsgesellschaft, Germany.

Part 4 for this book covers the key components of a Quality System. Three major parts of a Quality System are Auditing, Validation, and Supplier Qualification Systems. Chapter 9 describes how to audit a Quality System. Auditing is perhaps the most important tool to keep a system running and updated. In order to plan, carry out and control validation activities, a concept has to be implemented which assures consistency between regulatory requirements and the realisation within the company. The Validation Master Plan represents the interface between System Quality and process quality. Chapter 10 covers this aspect.

Due to the rapidly changing pharmaceutical environment, the suppliers are playing a more important part. Strategies like outsourcing and business reengineering make it necessary to transfer tasks, which formerly have been part of the internal process, to a supplier. The Quality System, therefore, has to guarantee that the supplier is able to deliver the required quality.

A new trend in Quality System will be described in part 5 of this book. Hazard Analysis Critical Control Point (HACCP) is a tool in Quality Systems already used in other branches, for example, the food industry. But this tool is also applicable to the pharmaceutical industry. The last chapter describes the application of the new technique to the process validation in a sterile manufacturing company.

SUMMARY

A "golden key" for Quality Systems doesn't exist! But this book will give you advice and show you implementation reviews written by authors from different companies. Find your own way! Perhaps you conclude that it isn't that bad to have no exact guidelines to explain how to install and update such a system.

Oliver Schmidt

Acknowledgments

This book wouldn't be possible without the work of my coauthors. I want to express special thanks to my colleagues. They have written their articles in their spare time, of which they don't have a lot, which is mostly taken up with their business activities:

Dr. Andreas Brutsche
Dr. Fritz Demmer
Dr. Lothar Hartmann
Dr. Michael Jahnke
Karl Metzger
Dr. Heinrich Prinz
Dr. Wolfgang Schumacher
David Sullivan
Rudolf Völler

Author Biographies

Oliver Schmidt, Editor, completed studies at the University for Applied Studies in Ludwigshafen and Worms, Germany. He has been the head of GMP Training and Consulting and a partner in Concept Heidelberg since 1996. Mr. Schmidt is a qualified ISO 9000 auditor and a qualified environmental system auditor according to the German Society of Quality regulations. He is also certified in EN ISO 10011, part 2.

Andreas Brutsche is Head of the Compliance Department in Quality Operations for Novartis Pharma AG, Basel. He is a European Union Qualified Person and is head of the expert group Quality Assurance of the Association of Pharmaceutical Technology.

Fritz Demmer has 26 years of experience in the pharmaceutical industry in various management functions including production, development and quality assurance. Dr. Demmer has been a consultant for the pharmaceutical industry in all aspects of QA including validation since 1995.

Michael Jahnke has a PhD from the Institute of Microbiology at the University of Hanover. He was the Head of the Biology Department at IBR Forschungs GMBH and is currently the Head of the Quality Control/Microbiology Department at Pharma Hameln. Dr. Jahnke is a member of the American Society for Microbiology

(ASM) and of the Parenteral Society. He is on the editorial board of the European Journal of Parenteral Sciences.

Lothar Hartmann has a PhD from the Technical University in Berlin. He worked as a plant manager for Hoffmann-Laroche in Germany before moving to Hoffmann's headquarters in Switzerland. He is currently responsible for GMP Compliance and Auditing for the Vitamins and Fine Chemicals Division of Roche. Dr. Hartmann is the head of the working party GMP and Quality Systems for CEFIC's BPC Committee and has had an article published in a joint CEFIC-EFPIA journal.

Karl Metzger has a degree in chemistry from the University Kaiserslautern. After working in various research institutes for several years he joined BASF Pharma where he was GMP/DMF Coordinator at the Uetersen plant of Knoll AG and member of BASF Pharma's Validation Group. Mr. Metzger has been a project manager at Concept Heidelberg since 1997. He is responsible for training and consulting projects regarding pharmaceutical starting materials, sterile manufacturing, and R&D.

Heinrich Prinz is Head of the Quality Assurance Department at Biotest AG where he is responsible for the overall uniform Quality Assurance System in all the company's pharmaceutical fields throughout Europe based on legal requirements, PharmBetrV, EC-GMP Directive and Guideline, and CFR, DIN EN ISO 9001. He is a certified auditor and, with Det Norske Veritas, runs certification audits for quality systems and CE marked for medical devices based on European regulations.

Wolfgang Schumacher qualified in pharmacy and organic chemistry at the University of Heidelberg. He did his thesis at the Max-Planck Institute for medical research and then joined the research department of ASTA Medica where he specialized in anticancer drug discovery. After heading a European harmonization project he has been head of GMP Compliance International at ASTA's Frankfort headquarters.

David Sullivan, who has an MS in chemistry from University College, Cork, Ireland, spent 27 years at Hoechst Marion Roussel in various positions, including site QA/QC manager in Canada and France. More recently he coordinated and led a global quality team responsible for corporate management groups in QA, auditing, international standards, microbiology and packaging graphic standards at Hoechst's Frankfurt headquarters. Mr. Sullivan currently lives in Brussels and works as a consultant.

Rudolph Völler completed studies in pharmacy and food chemistry in Frankfort. He is a member of the administrative pharmaceutical department at the Regierungspräsidium in Darmstadt where he is responsible for inspections of pharmacies, hospital pharmacies, pharmaceutical mnufacturers (GMP, laboratories (GAP), and clinical trials (GCP). He has performed foreign inspections in Korea, Saudi Arabia, the United States, and Canada and is a member of expert groups in quality and inspections.

1

Introduction in Quality Systems

Oliver Schmidt
Concept Heidelberg
Heidelberg, Germany

Rudolf Völler
Regierungspräsidium
Darmstadt, Germany

A book entitled *Quality Systems* must contain, of course, some definitions of terms. Therefore, we want to begin with a few basic explanations. To find the correct definition of a quality system, you first need to understand what is meant by quality. However, even for this very first term there is no standard definition. Unlike natural science, where only clearly defined terms are permitted, the whole subject area of quality, quality assurance (QA), and quality management is not easily described nor defined.

WHAT IS QUALITY?

Meyer's Lexikon (1993) defines "quality" as character or nature, grade and value. Ever since Aristotle, quality has been a category classifying evaluations of essential properties of objects, usually perceivable by means of the senses.

If you take the German Medicinal Products Act, quality characteristics mentioned in §14 (15) are identity, content, and other properties. However, there is no generally accepted definition of this term.

WHAT IS QUALITY ASSURANCE (QA)?

QA is nothing new in pharmaceutical manufacturing. However, how to define and distinguish among QA, quality control (QC) and Good Manufacturing Practice (GMP) has been a recurring issue for discussion. In the 1989 European Commission (EC) Guide to Good Drug Manufacturing Practice, the first chapter is dedicated to QA (strangely enough under the heading "Quality Management"). There you can read: The pharmaceutical manufacturer shall warrant the quality of the drugs. "To safeguard that this is achieved, the manufacturer must have a comprehensively designed and correctly implemented QA System that includes GMP and thus QC. This system should be completely documented and its efficacy monitored."

1.1 The basic concepts of QA, GMP and QC are interrelated. They are described here in order to emphasise their relationships and their fundamental importance to the production and control of medicinal products.

1.2 QA is a wide ranging concept covering all matters which individually or collectively influence the quality of a product. It is the sum total of the organised arrangements made with the objective of ensuring that medicinal products are of the quality required for their intended use. QA therefore incorporates GMP plus other factors outside the scope of this guide.

1.3 GMP is that part of quality assurance which ensures that products are consistently produced and controlled to the quality standards appropriate to their intended use and as required by marketing authorisation or product specification. This makes it clear that quality assurance is the generic, wider term. GMP is just one component of quality assurance, albeit the essential one. Classic quality control deals mostly with the required inspections and is governed by the GMP rules. The terms "quality assurance" and "quality control" should be clearly distinguished. GMP is a part of quality assurance, but quality assurance is more than just GMP (see Figure 1.1).

WHAT IS QUALITY MANAGEMENT?

The term "quality management" is closely related to the term "quality assurance" and can be explained as follows: QA includes any and all social and technical measures taken to ensure a standard quality of the production process turnout. In terms of organisational lines, quality management means the individuals in a corporate organisation who are in charge of QA. By this definition, quality management would thus be a department. By the term "quality management system" we mean the organisational structure, the assignment of responsibilities and powers, procedures and processes, as well as the resources required to achieve quality.

Another definition is given in the 1992 World Health Organization (WHO) GMP Guide. Although the WHO GMP Guide is not of foremost authority, the definition given there is a rather good one:

> In the drug industry at large, quality management is defined as the aspect of management function that determines and implements the "quality policy", i.e., the overall intentions and direction of an organisation regarding quality, as formally expressed and authorised by top management.

The basic elements of quality management are:

- an appropriate infrastructure or "quality system", encompassing the organisational structure, procedures, processes, and resources; and

- systematic actions necessary to ensure adequate confidence that a product (or service) will satisfy given requirements for quality. The totality of these actions is termed "quality assurance".

It is interesting to see that this definition leans on the definition contained in ISO 9000.

Figure 1.1. Definitions.

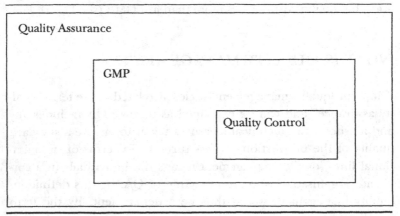

WHAT IS A QUALITY SYSTEM?

To come back to the definition given in the WHO's GMP, a quality system is a component of quality management, "encompassing the organisational structure, procedures, processes and resources; and systematic actions necessary to ensure adequate confidence that a product (or service) will satisfy given requirements for quality".

That is to say, in order to ensure quality, there must be an encompassing system. This includes the requirement that a philosophy of quality be drawn up. The European Agency for the Evaluation of Medicinal Products (EMEA) document contains a particularly neat philosophy of quality:

> Quality needs to be built from the beginning. Independent of any quality system, any old or new management style, any time or location, is the plain statement that quality cannot be inspected or audited into a process, a system, an organisation. Quality needs to be built in, in every little task and service, by everyone participating in the process. The need to do so starts at the design phase of the product, which can be a car, a piece of furniture, a new production line, a whole factory or an organisation created to provide a service.

What Are the Benefits of a Quality System?

The EMEA also offers a very good explanation of the benefits of quality systems:

> Sceptical persons may be in doubt about any benefit and on the contrary point out that the time spent on the QMS exercise could have been better used, to produce more and faster. No bridge is ever built without logistics, no battle won without a strategy, no tree grew, blossomed and gave excellent fruit without care, cutting and guiding branches where necessary. Potemkin's village, where decorated facades were hiding misery, should not serve as an example.

How to Present a Quality System

The answer to this question seems to be simple: With appropriate documentation! However, this gives rise to the next question: What is appropriate documentation? If there was an easy answer to this question, you would not be reading this book.

If there are no generally accepted definitions of terms, and current opinions on how quality systems should be documented diverge even more widely. This book gives you practical advice and interpretations by consultants, QA professionals of major and medium-sized companies, and a GMP inspector as a basis for developing your own documentation, tailored to your specific business situation.

Definitions

The revision of the ISO-9000 series also involves the revision of the definitions contained in these standards. In November 1998, the German Institute for Standardisation (DIN) published the first draft of the ISO 9000, containing the following definitions:

- System—Interrelated and interacting processes working in harmony.

- Management System—System to establish and fulfil the policy and objectives of an organisation.

- Quality Management System (QMS)—Part of a management system related to the establishment and fulfilment of quality policy and quality objectives.

- Quality Planning—Part of a quality management system focused on establishing and/or interpreting quality policy, quality objectives, quality targets, quality requirements, and defining how these are to be achieved.

- Quality Control—Part of a quality management system focused on the operational techniques and processes used to fulfil quality requirements.

- Quality Assurance—Part of a quality management system focused on the fulfilment of quality requirements and providing confidence of meeting customer requirements.

A collocation of these definitions leads to the graph found in Figure 1.2 (designed by the authors).

It is obvious that there are differences between the definitions of ISO 9000 and GMP, especially in terms of QA and QC. According to the GMP Guide, QC was part of GMP, and therefore part of QA. We believe that the ISO 9000 definition is more logical and will become a worldwide standard. However, harmonisation is demanded, irrespective how these common standard will look. This will help to avoid future misunderstandings.

THE DOCUMENTED QUALITY SYSTEM

In the following, we want to present a typical documentation structure as it is used by many companies. Naturally, any other procedure may be just as good since there are no statutory definitions, let alone provisions for documenting a quality system, as has been shown in Figure 1.2. Accordingly, a quality system is structured as can be seen in Figure 1.3.

The quality manual is at the top of the pyramid. A quality manual should be written for several reasons.

- The authority supports the establishment and implementation of an effective quality assurance system which should be completely documented (Chapter 1, EC GMP Guide). The authorities will increasingly monitor the fulfilment of these requirements and demand that a quality manual be submitted.

- In a quality manual, all measures designed to ensure quality are compiled and clearly presented in their entirety. This form of documentation shall serve internal verification purposes and make sure that good manufacturing practices are employed fully in a company. Also, the quality manual is a guideline for self-inspection to ensure compliance with GMP rules established for in-house purposes.

- Anyone who thinks that a quality manual serves to set out instructions for the production and quality control staff,

Figure 1.2. Definition of terms.

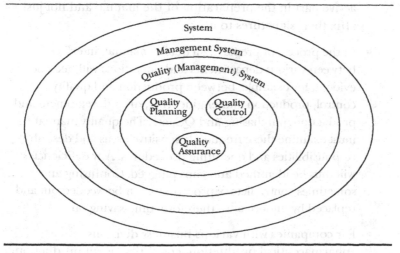

Figure 1.3. Documentation of a quality system.

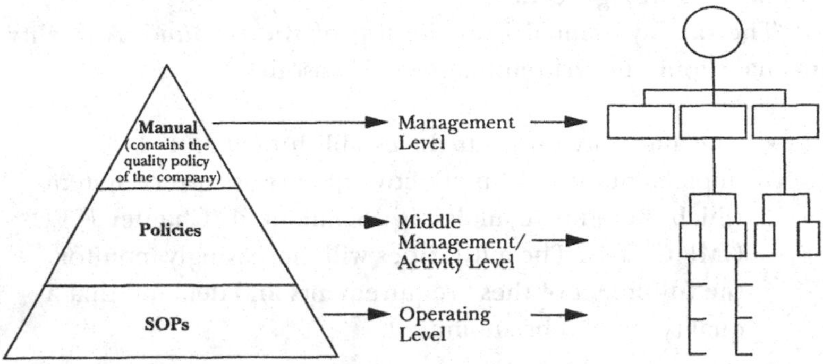

has failed to understand one important aspect of quality manuals. Quality production is an element of a company's policy and thus of its corporate identity. The management's quality policy stated expressly in the quality manual ensures the employees' understanding and support of measures to improve quality. A crucial factor in motivating a workforce to produce good quality is the stated position of the top management on this issue. Therefore, competencies and responsibilities are of particular importance in the context of a quality system. This first section in a quality manual should be taken seriously, and the individuals in charge should take an active part in the preparation of the manual and not just affix their signatures to it.

- In the process of preparing a quality manual, interfaces between various departments and operations will become evident, for example, between production and quality control, production and engineering, drug development and production, purchasing and QC, etc. The quality manual must examine these frequently sensitive areas and describe responsibilities and operating procedures. Any deficiencies will thus be identified and gaps plugged. Confusing and sometimes contradictory procedures can be weeded out and replaced by smooth and, therefore, time-saving ones.

- For companies with various business divisions (pharmaceutical production, cosmetics, agent production, diagnostics, etc.) or plants at various sites, the quality manual is an ideal tool to exchange information and prevent that the same jobs are carried out twice, for example, the preparation of guidelines. The harmonisation of the various areas and their integration under the umbrella of a uniform and representative quality assurance is a welcome effect.

- Companies receive more and more requests from their business partners for information on their quality system concept. For suppliers of active ingredients, packing materials, or medicinal products, it already goes without saying that they have their quality manuals they can submit for auditing purposes. Subcontractors, too, are well advised to have a representative quality manual they can present.

You will find various examples of possible manual structures in this Interpharm book.

At the next level below the top, there are the policies. These types of documentation are found primarily in major companies. QMSs are usually managed centrally and describe cross-division operating procedures, for example, validation policy, documentation of in-house training activities, and validation principles of computer-assisted systems.

Policies are always useful where company-wide standard rules are preferable to a host of individually established rules, which may sometimes be contradictory. Policies frequently contain descriptions of interfaces, i.e., they explain business processes involving 2 or more departments. A good example is the purchasing of starting materials. In many companies, the purchasing department is separate from the actual quality control department. It must be ensured, however, that the purchasing staff place their orders only with approved suppliers at the given specifications. The policies are usually steered from a central, cross-departmental QA department. Therefore, policies are well-suited to define interfaces between purchasing, like in this example, and quality control.

The third level are the standard operating procedures (SOPs). SOPs contain instructions for the workforce on how to handle production-related procedures. In the ISO 9000, the term quality system procedures (QSP) is used for this, whereas the ECGMP Guide refers to it as "procedures".

According to the glossary of the EC GMP Guide, procedures include the *"description of operations to be carried out, the precautions to be taken and measures to be applied directly or indirectly related to the manufacture of a medicinal product"*.

Paragraph 4.1. of the chapter on documentation reads: *"Procedures give directions for performing certain operations, e.g., cleaning, dress code, environmental control, sampling, testing, equipment operation"*.

In contrast to documentation levels 1 (manual) and 2 (policies), the EC GMP Guide addresses SOPs in concrete terms. In chapter 4, SOPs are expressly required on the following issues:

- Acceptance of starting materials and packing materials (goods receiving)

- Marking, quarantine and storage of starting materials and packing materials

- Sampling

- Inspection of materials and products at the various production levels

- Clearance and rejection of materials and products

- Validation

- Assembly and calibration of equipment

- Maintenance, cleaning and disinfection

- Personnel-related issues (training, changing of clothes, hygiene, etc.)

- Environmental monitoring

- Pest control

- Complaints, recall actions

- Returns

- Clear operating instructions on essential parts of production and inspection equipment

The above mentioned SOPs are expressly required. Naturally, this list is incomplete. For warehouse management alone, for example, the following SOPs can be deduced from Good Storage Practices:

- Creation of purchase requisition data records

- Goods receiving

- Shipping

- Returns

- Clearance, qualified clearance, blocking

- Complaints

- Storage area for blocked stock

- Storage area for advertising material

- Hygiene plan

- Pest control

- Waste disposal

- Hazardous materials handling

- Monitoring of weighing instruments

- Temperature and humidity monitoring

- Warehousing and distribution by subcontractors

SOPs play an important role in quality systems. They serve to implement the principles laid down in the quality manual. Being a flexible, adjustable part of the documentation, they guarantee reproducible business processes.

When preparing SOPs, care should be taken not to describe product-specific processes. Product-specific processes, for example, for validating, manufacturing or packaging, should not be laid down in SOPs but in the respective manufacturing or validation rules. The following principles apply to SOPs:

1. They shall contain a detailed description of procedures and processes.

2. They shall reflect reality.

3. They shall be written in a way that is easy to understand.

4. They shall have a uniform layout.

5. They shall be identified by means of a numbering system.

In view of the large number of SOPs issued in a company, a boilerplate SOP should be written to ensure standardisation in the various fields of business processes (production, quality control, packing, warehousing, etc.). It is not infrequent that boilerplate SOPs are deficient insofar as important directions that need to be laid down are omitted. The following checklist has been designed to ascertain that all required directions have been included in a boilerplate SOP:

1. Form

   • Which form layout (header/footer, font size, etc.) is to be used?

2. Author

   • Who is responsible for writing an SOP?

3. Verifier

   • Who is responsible for verifying an SOP?

4. Putting into force

   • Who puts an SOP into force and when will the SOP come into effect?

5. Distribution (including identification of the original document)

   • Who distributes the SOP and what is the actual procedure?

6. Amendment/revision

   • Who is responsible for amending/revising SOPs, and at what point in time?

   • What is the procedure for revising and replacing outdated SOPs?

7. Declaration of invalidity/withdrawal of invalid SOPs/disposal

   • What is the procedure for declaring an SOP invalid and withdrawing its copies?

8. Deviations/changes

   • What is the procedure in case of a deviation from an SOP direction?

9. Numbering and version number

   • What is the alphabetical/numerical code used to identify SOPs and how is the version number shown?

10. Archiving

    • Who is responsible for archiving SOPs, in which location and for how long?

11. Structure

    • How should each SOP be structured?

12. Annexes/jointly applicable documents

    • How are annexes and jointly applicable documents to be handled?

13. Training

    • How are new subject matters of SOPs imparted to the staff and how is their training documented?

CONCLUSION

The problem is that there is currently no generally accepted definition of the term "quality system" that would clearly distinguish it from other terms used in this context. The GMP Guide of the EC contributes to the confusion insofar as its first chapter is entitled "Quality Management", but in the following, only the term "quality assurance system" is used. Because the new definition given in ISO 9000, which was mentioned previously, is easy to understand, we hope that it will gain acceptance.

Until this happens, each company will have to lay down its own definitions and introduce its own rules governing the documentation of its quality system. See the following chapters for various examples.

2

Experiences in the Inspecting of Quality Systems— An Inspector's View

Rudolf Völler
Regierungspräsidium
Darmstadt, Germany

REGULATORY REQUIREMENTS

Quality Assurance (QA) systems are the statutory norm for pharmaceutical entrepreneurs in Europe. The individual inputs are either made binding by the European Economic Community (EEC) regulations or they were translated into national standards in the European Union (EU) Member States in accordance with the corresponding EEC Directives (Figure 2.1).

The European Commission (EC) Directive identifies the inputs required by the EU Guideline as state-of-the-art (Table 2.1).

QA systems are described comprehensively in this guideline (Table 2.2).

Figure 2.1. Implementation of QA and GMPs in Europe.

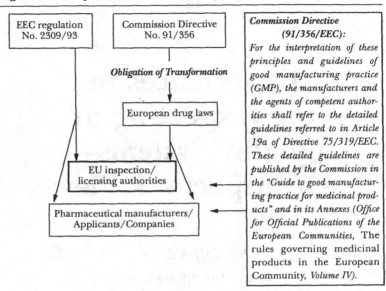

Table 2.1. EC Directive on Quality Assurance.

EEC-Guide: The basic concepts of QA, GMP, and QC are interrelated. They are described here in order to emphasise their relationships and their fundamental importance to the production and control of medicinal products.

Quality Assurance: QA is a wide-ranging concept that covers all matters which individually or collectively influence the quality of a product. It is the sum total of the organised arrangements made with the object of ensuring that medicinal products are of the quality required for their intended use. QA therefore incorporates good manufacturing practice, plus other factors outside the scope of this guide.

Table 2.2. Guideline for QA Systems.

EEC-Guide: The system of QA appropriate for the manufacture of medicinal products should ensure that:

1. Medicinal products are designed and developed in a way that takes account of the requirements of good manufacturing practice and good laboratory practice.

2. Production and control operations are clearly specified and good manufacturing practice adopted.

3. Managerial responsibilities are clearly specified.

4. Arrangements are made for the manufacture, supply and use of the correct starting and packaging materials.

5. All necessary controls on intermediate products and any other in-process controls and validations are carried out.

6. The finished product is correctly processed and checked according to the defined procedures.

7. Medicinal products are not sold or supplied before a qualified person has certified that each production batch has been produced and controlled in accordance with the requirements of the marketing authorisation and any other regulations relevant to the production, control and release of medicinal products.

8. Satisfactory arrangements exist to ensure, as far as possible, that the medicinal products are stored, distributed and subsequently handled so that quality is maintained throughout their shelf life.

9. There is a procedure for self-inspection and/or quality audit which regularly appraises the effectiveness and applicability of the quality assurance system.

It is clear that the QA of all relevant areas is taken into account in manufacture and marketing authorisation. This general requirement must be implemented in the laws and ordinances of the individual countries. In Germany, the Operation Ordinance for Pharmaceutical Entrepreneurs (PharmBetrVO), for instance, requires that factories and installations operate a functioning pharmaceutical quality assurance system.

QA systems are therefore objects of inspection by the authorities and may be enforced by means of impositions. The absence of adequate systems may also lead to the suspension or revocation of a licence, since they cannot be ensured without GMP-compliant manufacture. German drug law contains the obligation of the supervisory authority to suspend or revoke the licence "if the manufacturer is not in a position to ensure that the manufacture or testing of the medicinal product is carried out according to the state of the art".[1]

MANAGEMENT REPRESENTATIVE OR QA DEPARTMENT?

Involvement of the Management

QA systems must involve the management so that relevant decisions can be made and the final responsibility for the quality of the medicinal product and its marketing can be realised. The QA system is therefore the link between the individual key persons, the responsible persons, other decision makers, and the management (Figure 2.2).

The QA system must combine the individual parts and enable and ensure the assumption of responsibility at the individual positions in a company.

This quality structure must be described. Both the system and the corresponding measures are to be documented.

Example:
The Commissioner for the Graduated Plan must record events and coordinate the necessary measures. Any problems

---

[1]This regulation became effective on 1 April 1999.

Figure 2.2. QA system and background.

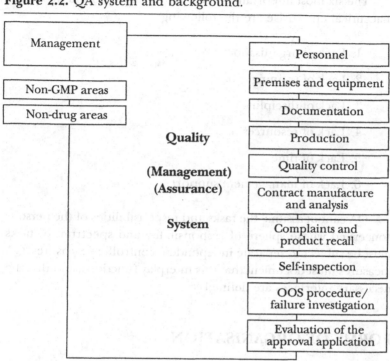

in manufacture or quality control must be communicated to the respective responsible persons. The communication system necessary for this must be structured and functional, that is, be subject to controlling measures.

This implementation of the entire system is not the task of the individual responsible person. Rather, it is the task of the pharmaceutical entrepreneur, or of the respective management whose involvement is already ensured by means of general laws (e.g., norms of commercial or criminal law). This requirement, however, is also part of the relevant norms of drug law. It is also the task of the individual persons to examine the efficacy of the system in their own sphere of influence for its plausibility, and to point out defects and faults in writing.

The six most important defects which can be found in a pharmaceutical enterprise are the following:

1. Lack of organisation

2. Lack of training

3. Lack of discipline

4. Lack of resources

5. Lack of time

6. Lack of management support

QA cannot assume the tasks and responsibilities of the persons concerned whose sphere of responsibility and spectrum of tasks must be subject to separate independent controlling measures. QA measures must document that this interplay functions, and that the necessary interfaces are defined.

FORMS OF ORGANISATION

The type of QA system and its form of organisation are not expressly stipulated.

First of all, it must exist in writing and the measures must be documented. This complex job cannot take place without assigning responsibility, or without a description of the tasks involved. The tasks must be assigned to certain persons who implement the measures, evaluate and report the results and enable corrections to be made.

It is interesting that PharmBetrVO provides for self-inspections which must be independent, meaning that the QA activities are also subject to an internal audit.

Additional QA measures therefore require additional personnel support. This work is not contained in the job descriptions of the other responsible persons and must be formulated separately and charged to certain persons. It is therefore advisable to select independent persons, but German law does not require that independent persons be assigned.

National and international regulations specify that sufficient and adequately qualified persons must be on hand for the jobs and tasks arising. As a rule, this means that one or several persons must be expressly commissioned for the implementation of the QA system. These people occupy themselves with implementing the task, and they maintain and update the system after its introduction. Larger enterprises have separate QA departments for this task. The relevant papers describing the system and the records of the action taken must be on hand.

The basic idea of implementation of a QA system depends on the independent assessment of the manufacturing processes by institutions not involved in manufacture. It is an interesting fact that in Germany a divided responsibility for the manufacture of medicinal products has been required by law since 1976: production and testing must be the responsibility of independent qualified persons.

This consistent separation of manufacture from control often does not exist in other countries, which explains the special emphasis on the importance of an independent QA system. In the United States, an independent body must perform a documentation review. Subsequent inspection of the papers, therefore, serves the purpose of ensuring the compliance of the processes and all accompanying circumstances in production and QA.

The systems are therefore the same; manufacture and testing of the product must be independent of one another. The system must also review all additional decisive parameters at regular intervals. The task of a QA system is both to ensure the system in the manufacture and testing, and to ensure that all processes comply.

By definition, however, all quality-relevant areas and also the areas of the organisation up to the highest level must be taken into account, and this is where self-inspection acquires the character of a management review.

THE DIFFERENCE BETWEEN THE UNITED STATES AND EUROPE

The difference in viewpoint between Europe and the United States appears to be particularly problematic at present, when one considers the incorporation and task structure of the QA system and the implementation and responsibility of in-process control and release.

Some interpretations and demands of the Food and Drug Administration (FDA) inspectors include the following:

- It is interesting to note that the FDA regards the combination of QC and QA as a single unit. Organisational regulations with a division of tasks and some degree of independence from one another appear to be of subordinate importance. (See Table 2.3.)

- The QC/QA department is also responsible, in the opinion of the FDA, for the perusal of all documents; any soft data such as environmental controls and monitoring reports are also to be included in the release decision.

Table 2.3. The FDA statements on QC and QA.

---

21 CFR 211.22:
There shall be a quality control unit that shall have the responsibility and authority to approve or reject all components, drug products, containers, closures, in-process materials, packaging material, labeling, and drug products, and the authority to review production records to assure that no errors have occurred, that they have been fully investigated. The quality control unit shall be responsible for approving or rejecting drug products manufactured, processed, packed, or held under contract by another company.

Discussion paper of the FDA on the topic of cGMP (May 1996) on the topic of quality control and quality assurance concerning 21 CFR 211.22:

The FDA does not believe that such changes in terminology would be useful. The difference between quality assurance and quality control is recognized to be operational. The quality control unit is usually responsible for performing the testing to assure that proper specifications and limits are adhered to, while the quality assurance unit is responsible for auditing methods, results, systems and processes and for performed trend analysis.

> Remark: The European regulations also make
> provision for this inclusion of additional data.
> Although the American authority has recently
> expressed a different opinion, a fundamental
> involvement of quality control in the responsibil-
> ity in the case of the implementation of the mea-
> sures cannot be deduced from the Code of
> Federal Regulations (CFR.) (See Table 2.4.)

It is expressly expected by the FDA that QC/QA peruse the manufacturing data so that the necessary findings, for example, the deviations in the manufacturing process in the case of analysis, are known and can be considered accordingly (important in the case of out-of-specification (OOS)/error evaluation/revalidation).

- Any changes and the analysis of the in process control (IPC) connected with this must take place with the knowledge of the QC/QA unit; the analysis is to be validated by QC/QA.
- Quality inspections and tests during manufacture and IPC are to be carried out by the QC and QA department (This cannot be absolutely deduced from the CFR. See Tables 2.3, 2.3, 2.5).

The demands of the American colleagues are for the large part logical and understandable. They do, however, go beyond ideas of the European view of GMP and appear at first glance to be incompatible with some demands of the European regulations.

If one applies the standards valid in Europe to this interpre-tation, first of all the text of some regulations should be cited:

Table 2.4. Discussion paper of the FDA on the topic of GMP for the Manufacture of Active Substances (Draft 1995).

Some tests may be conducted by qualified production department per-sonnel and the process adjusted without prior quality control approval provided adjustments are made within limits preestablished and approved by the quality control unit.

Table 2.5. 21 CFR 211.110 (c).

In-process materials shall be tested for identity, strength, quality, and purity, *as appropriate*, and approved or rejected by the quality control unit, during the production process, e.g., at commencement or completion of significant phases or after storage for long periods.

The modification suggestion (1996) comments:
The regulation is designed to protect the integrity of the manufacturing process.

Chapter 1 EC/International Inspection Convention (PIC)-GMP Guideline: Quality Management (QM)

To achieve the quality objective reliably there must be a comprehensively designed and correctly implemented system of Quality Assurance incorporating Good Manufacturing Practice and thus Quality Control. All parts of the QA system should be adequately resourced with competent personnel, and suitable and sufficient premises, equipment and facilities. QA therefore incorporates GMP plus other factors outside the scope of this Guide.

The system of QA appropriate for the manufacture of medicinal products should ensure that all necessary controls on intermediate products, and any other in-process controls and validations are carried out [and] there is a procedure for self-inspection and/or quality audit which regularly appraises the effectiveness and applicability of the QA system.

The basic requirements of QC state that product assessment includes a review and evaluation of relevant production documentation and an assessment of deviations from specified procedures.

Chapter 9 EC/PIC-GMP Guideline

Self-inspections should be conducted in an independent and detailed way by designated competent person(s) from the company. Independent audits by external experts may also be useful.

6.17 EC/PIC-GMP Guideline

The tests performed should be recorded and the records should include at least a clear statement of release or rejection.

And finally, a quote from the German Operation Ordinance for Pharmaceutical Entrepreneurs (PharmBetrVO):

> § 7 Release
> (1) Medicinal products may be only marked as released (release) if the manufacturing and test records are duly signed.

As a rule, the requirement of independence of the QA unit is derived from the circumstance that this is responsible also for control measures conducted in the form of self inspections and must take into account the company as a whole. For this reason, QA is often understood as a staff position independent of QC and directly responsible to the management.

In my opinion this is still the most favourable solution.

On the other hand, at least according to the EC GMP Guideline, QC is involved in the inspection and control of all manufacturing data. The guideline provides for an individual inspection of the documents (EC GMP Guideline No. 6.3).

The European standards do not prohibit an organisational linking of QC and QA systems. It is also acceptable if independent QA personnel inspect the entire documentation of the production and then makes a company-specific release decision (this process is of course documented). In the case of this construction it should, however, be ensured that the self-inspections are carried out by persons independent of the release decision. This must be solved organisationally in the QA department. This is specified and documented by job descriptions and organisation charts.

For instance, persons from the QA department who are involved in review and in the release decisions should not be charged with the self-inspection of these processes.

The American view can therefore be integrated into the European standards without difficulty. If one considers the illustration in Figure 2.3, this means that areas b, c and d can be consolidated.

Within this area, however, the following requirements must be ensured (interface description):

- The person who issues the release in QA is independent of QC (independent if possible in the case of the European interpretation; absolutely independent according to the FDA).

- The persons who carry out the self-inspections must be independent from QC and the person who releases the batches.

The solution that is establishing itself in large companies consists of a clearly structured QC/QA department in which the respectively necessary independencies are defined by interface description.

It should also be pointed out that, according to German law, release results automatically when the manufacturing record and test record are duly signed. The problematics of this regulation are not the subject of discussion here. The PharmBetrVO does not regulate how release for marketing is implemented. The internal standard operating procedures (SOPs), however, must be precisely defined.

Figure 2.3. Elements of quality management.

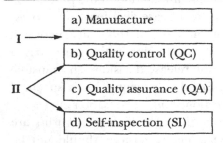

I. **Clear separation** of the implementation of the individual tasks.
II. **Interface** with clear description of the activities and responsibilities.

EVALUATION OF QUALITY SYSTEMS BY THE SUPERVISORY AUTHORITY

The Site Master File (SMF)

This interplay must be taken into consideration for the risk-graduated evaluation of an enterprise by the authority. The basis of such an evaluation is, as a rule, an SMF in which the system is adequately described (Table 2.5).

The supervisory guidelines in Germany specify that the inspectorates should have an SMF of the company at their disposal in order to be able to prepare for the inspections. As a rule, this documentation of the company is presented at short notice, although the presentation of this document is not regulated by binding norms. Since the company description is indispensable for a QA system, the inspectorate's demand can be justified according to such an SMF. A pharmaceutical enterprise cannot operate a functioning QA system if it does not have a company description on hand which corresponds to a SMF. It is interesting that the PIC SMF does not expressly specify the QA system, but the enterprise should comment on it in the chapter on self-inspections. In Europe the PIC SMF is currently in the testing phase, and the experience to date has been good. The question of the updating modalities should be agreed upon with the responsible authority in each case. As a rule, it is expected that minor changes

---

Table 2.5. Content of PIC SMF.

1.1 General information

1.2 Personnel

1.3 Premises

1.4 Documentation

1.5 Production

1.6 Quality control

1.7 Distribution

1.8 Self-inspection

---

be communicated in good time and the inspectors have the current version when the inspection begins. The SMF offers a common foundation for the GMP inspections.

The QA system is specified as an item to be evaluated in the European inspection report.

The evaluation of a QA system always takes place before the background of all processes in the manufacture of medicinal products.

Important shortcomings in sub-areas always indicate a non-functioning system. Too many defects in the details place the QA system as a whole in question, regardless of how it is structured. In an inspection, the evaluation of the QA system is not concluded with the review of the QA department. It is only after the evaluation of the entire system that it is determined whether the QA activities are effective. The evaluation of an enterprise as a whole cannot therefore be restricted to the review of the QA department.

Evaluation Criteria for a "Functioning" Quality System

In the manufacture and testing of medicinal products, the responsibilities of individual persons are clearly assigned, and interfaces and unregulated areas cannot be avoided. In order to ensure a manufacture in conformity with marketing authorisation and specifications, the entrepreneur must operate a QA system which covers all areas. This system also includes measures and criteria which are not assigned expressly to the spheres of responsibility of the respective responsible person(s). Various measures, such as environmental controls and systematic evaluation in connection with the batch records, are just as important as the compilation and evaluation of annual reports and holistic risk assessment of all activities; in the case of outsourcing, this expressly includes the contract acceptors.

The principle of a comprehensive batch record review is an essential component of this QA system. This system must be reviewed regularly by means of independent self-inspections. The entire QA system is consequently subject to inspections by the authorities. An inspection by an authority must take into account all quality-relevant areas of a company. Conclusions can be drawn, and they are an integral part of risk-graduated supervision.

In Europe the inspectorates are charged with these supervision measures. In Germany the tasks and powers of authority are based on § 64 of the German Drug Law (AMG). The result and the evaluation of an inspection are recorded in an inspection report. The European inspection report was adopted by the German inspection authorities and provides for comments and remarks on the following aspects (Table 2.6).

In particular, items 4, 8, 9 and 10 contain information relevant to a QA system. An inspection must evaluate both the system (see Figure 2.4) and the compliance with the specified inputs (see Figure 2.5). An inspection has several tasks: it considers and evaluates the details evaluates observed defects, and evaluates the whole. Massive defects in individual areas of an enterprise must lead inspectors to conclude that the QA system as a whole does not function. The QA system is an umbrella for all activities of an enterprise.

In order to evaluate a QA system, various questions must be asked (see Table 2.7).

From the answers one can determine whether the structural organisation and the sequential organisation have been described systematically and functionally, as well as in agreement with actual circumstances. The latter aspect is of special significance; the answer to the question "why" or "in which way" must in any case be followed by the request "show me".

Particularly in the review of the QA system, the inspector has the opportunity to evaluate the operation as a whole at every position in the operation and in every process.

Hierarchical structure and job descriptions in which as little overlap as possible occurs generally require some advice to ensure the doubt-free assignment of tasks and responsibility. Frequently the channels of communication are not clear, and the question of whether the inspectorate should collect the information or the entrepreneur should bring it to the inspectorate is not clearly answered.

Unclearness concerning substitution and the way responsibility is passed on is often observed. An enterprise's hierarchy can be systematically illuminated by means of the following questions (Table 2.8).

The control of the processes and the connection with other manufacture and testing-relevant procedures are an important part of the holistic QA system. For instance, it must be described to what extent repair work on the machines and premises, environmental controls, health examination of persons and change of suppliers of starting materials have been taken into consideration in the evaluation of the manufacture and testing of the medicinal products. The corresponding reporting procedures, evaluation options of the responsible persons and a suitable decision must exist if the individual processes and the specification-compliant observation of the parameters are to be documented.

Increasing use of a systematic description and evaluation of the medicinal products, in the form of a history of product with an evaluation of the results and any deviations, must be made. The regular observation of the batches and the direct comparison over a prolonged period yield indications as to the safety of the process. Decisions in favour of revalidation measures cannot be seen here. Out-of-specification (OOS) results can be evaluated in context. One often forgets that in the case of outsourcing, the contract acceptor must be adequately integrated into the QA system.

Table 2.6. Content of EU inspection report.

1.	Quality management (system)
2.	Personnel
3.	Premises and equipment
4.	Documentation
5.	Production
6.	Quality control
7.	Contract manufacture and analysis
8.	Complaints and product recall
9.	Self-inspection
10.	OOS procedure/failure investigation
11.	Evaluation of the approval application

Figure 2.4. System inspection.

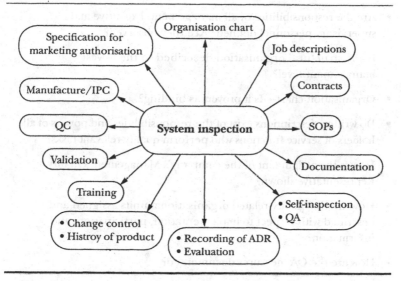

Figure 2.5. Compliance inspection.

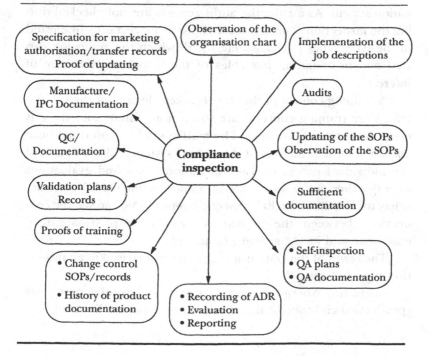

Table 2.7. Questions about the QA system.

- Are the responsibilities of all management, executive and supervisory personnel defined? Are they observed?

- Is the structural organisation described to the lowest management level?

- Organisation chart—Is it proven as binding?

- Do written descriptions exist of the responsibilities and powers of all holders of service functions who perform quality-relevant tasks?

- How is the assignment of the company's Management Quality Representative shown?

- How are the quality-related organisational units assigned and described with respect to implementation, participation, and information?

- How are the QA measures documented?

In addition to the audits, additional data, such as the quality of the purchased goods or services must flow into a systematic evaluation system. As a rule, the audit reports are not checked during the inspection. The inspector is, however, certainly interested in the systematics (checklist, etc.) with which the audits are carried out. In addition, timetables of the inspection visits are of interest.

A training control of the staff is a key element of the QA system. Here training concepts are often nonexistent, and there is often no systematic concept. The staff should be informed both about the basic principles of the overall system and the details of the individual processes. Familiarisation times and evaluations after this time are to be provided for and documented. If one considers the details of the EU inspection report, one can see the connections between the individual tasks of a pharmaceutical enterprise and its QA system (Table 2.9).

The result of the GMP inspection is summarised in an inspection report (Figure 2.6).

Interesting synergetic effects between the several departments (production and control, distribution, development, licensing . . .)

Table 2.8. Job description of persons in key positions.

Questions about the hierarchy
To whom does the person in question report?
Who reports to the person in question?
Who substitutes the person in question?
Whom does the person in question substitute?

Questions about the spectrum of tasks
What are the main tasks?
What are the secondary tasks?
To whom does the person in question report?
Who reports to the person in question?

Questions about the work in committees
According to what criteria are decisions made?
Composition of the Committee members belonging to the company?
To what extent do external experts participate?

---

and the QA activities of the site are obvious. An inspection, which shows lots of serious and a couple of critical observations must come to the conclusion that the QA system doesn't run correctly. This is independent of having presented well documented activities of QA department during inspection.

Last, but not least: as a rule, the QA department is involved in follow-up activities of the manufacturers, as the internal evaluation of the inspection report and the corrective actions.

CONCLUSION

In carrying out the statutory tasks of a supervisory authority, the in-company measures (audits/self-inspections) overlap with the external measures of the inspectorate. This leads to synergetic effects whose benefit for the GMP sector is not yet very marked. The First Party Audit Programme of FDA deals with the benefit of in-company self-controlling mechanisms. According to this, the systematic self-evaluation of an enterprise, which must of course be documented, is to reduce the inspection activities of the

Table 2.9. Connection between task fields and quality measures.

Sector	Quality System
Personnel	Initial education Qualification Requalification Health Honesty
Premises and equipment	Maintenance Monitoring Cleaning Qualification/calibration
Documentation	Batch documentation Logbooks Self-inspection protocols Audit protocols
Production	Specifications Validation/revalidation Accordance with marketing licence History of product
Quality control	Specifications Validation/revalidation Accordance with marketing licence History of products
Contract manufacture and analysis	Contracts Compliance with marketing licence Audits Subcontractors Product quality Change control
Complaints and product recall	Reporting of complaints Communication of all involved parties Corrective measures
Self-inspection	Reporting Competence Measures
OOS procedure/ failure investigation	Complaint detection procedure Complaint reporting Cooperation/communication
Evaluation of the approval application	Pre-clinical activities GMP-relevant procedures GCP/GMP interactions

Figure 2.6. Inspection report.

Seal	REGIERUNGSPRÄSIDIUM* DARMSTADT

Regierungspräsidium Darmstadt, D-64278 Darmstadt
<Telephone: 0049-6151-126242 # Telefax: 0049-6151-125789>

GMP—INSPECTION REPORT

Inspected site:

ADDRESS:
TELEPHONE:
FAX:
E-MAIL:
REMARKS:

ACTIVITIES:

Production of active pharmaceutical ingredients	❏	Import	❏
		Labor	❏
Production of finished products	❏	Batch control, release of batches	❏
Packaging	❏		

REMARKS:

DATE OF INSPECTION:

INSPECTORS: Name of inspectors

　　　　　　Name of experts/other inspectors
　　　　　　N/A ❏
　　　　　　Name of the competent authority(ies):

REMARKS:
1. Introduction
2. Scope of inspection (Short description of the inspection).
3. Inspected area(s) (Each inspected area should be specified).
　　3.1
4. Personnel met during the inspection (key personnel met, names, titles).
　　4.1　and others

Continued on the next page.

Regierungspräsidium office of the chief official of the administrative district

Continued from the previous page.

5.   Inspector team's findings:

5.1   Quality assurance (system)

5.2   Personnel

5.3   Premises and equipment

5.4   Documentation

5.5   Production

5.6   Quality control

5.7   Contract manufacture and analysis

5.8   Complaints and product recall

5.9   Self-inspection

5.10 OOS procedure/failure investigation

5.11 Evaluation of the approval application

6.   Miscellaneous

6.1   Samples taken:

6.2   Distribution:

6.3   Assessment of the Site Master File:

7.   List of observations:

7.1   Critical

7.2   Serious

7.3   Minor

8.   Recommendations:

9.   Summary and Conclusions:

Sign and date:

Representative of the company   *Inspector:*

Other experts/inspectors involved:

☐ *N/A*

Authority:

authority. This interactive system is not unknown in Germany, and it is being tested in the field of industrial and environmental law. The law on medicinal products is also based on self-controlling. The implementation of effective QA systems provides the opportunity to reduce the authorities' commitment and to perceptibly strengthen the enterprises' self-responsibility without loss of safety.

3

GMP/ISO Quality Systems for Drug Products Manufacturers

Oliver Schmidt
Concept Heidelberg
Heidelberg, Germany

ISO 9000: A CRITICAL ASSESSMENT

Hardly any document, apart from the universally feared the Food and Drug Administration (FDA) Guidelines, has caused so much controversy as the ISO 9000 standards series. In addition to being a critical analysis, the discussion has also often taken on emotional elements. During this, the proponents and the opponents of ISO 9000 have often ended up in a "clinch" from which there appeared to be no escape. The American business journalist Ronald Henkhoff once summarized his view of ISO 9000 by remarking that the ISO 9000 standards can be summarized in three words: documentation, documentation and more documentation (*Frankfurter Allgemeine Zeitung* 1994). Somebody once said that these principles also apply to the FDA. You will see that this circumstance does not, however, bring us any further.

Henkhoff is not alone in his critical opinion of ISO 9000. In the German magazine *Manager*, one could read the following in the August 1995 article "The big bluff": "Certification, according to ISO, is first and foremost a big money-maker. Improved quality? More likely a regression into dull bureaucracy". The author, a business consultant and author of various books on management, expresses the opinion of many. His thesis: "Whoever is deeply in favor of Total Quality Management (TQM) must be against ISO 9000".

Many of you know comparable articles in which, rightly or not, ISO 9000 is examined more than just critically. Let us look not only at the critical voices but also at the development ISO 9000 has made within the pharmaceutical industry. In Europe, ISO 9000 has received a varied reception. For instance, ISO 9000 has established itself more easily in the pharmaceutical industry in Britain than it has in Germany. In September 1991 an article appeared in the trade journal *Pharmind* entitled "Quality assurance in the pharmaceutical industry" (Die Pharmazeutische Industrie, September 1991). It represented for the first time a technically-based comparison of the ISO 9000 series with the requirements from pharmaceutical guidelines such as the European Commission (EC) Guide to Good Manufacturing Practice (GMP) Guide. In general, however, the opinion was presented that the ISO 9000 series, as intersectoral standards, are nothing new to the pharmaceutical industry, since these pharmaceutical guidelines cover all aspects of ISO 9000. Articles such as "ISO goes to Washington" by Paul Scicchitano, published in *Compliance Engineering European Edition* (Scicchitano 1997), demonstrate, however, that ISO 9000 has also found its way into the arch-conservative FDA—although, for the time being, only in the area of medical devices.

I would like to explain in this article the statement ISO 9000 makes, what options it offers, and how it can be applied. I would also like to explain where the limits of ISO 9000 are, for we have without a doubt a very special situation in the pharmaceutical industry. On the one hand, there exists a practice of pharmaceutical guidelines that has been established and lived by for years (first and foremost the GMP Guide), while on the other hand an internationally recognized body of quality standards exists.

HOW THE ISO 9000 SERIES CAME ABOUT

ISO 9001 defines itself in the section "Scope": "This International Standard specifies quality system requirements for use where a supplier's capability to design and supply conforming products needs to be demonstrated. The requirements specified are aimed primarily at achieving customer satisfaction by preventing non-conformity at all stages from design through to servicing".

Even after reading it through several times, you will notice that the meaning of this statement is not easy to grasp. As I understand it, the purpose of ISO 9000 is to ensure the quality of a product by means of a large number of measures. In order, therefore, to approach the content of this standard once again, we first want to define what quality actually is.

The various definitions range from "Quality is if it works", to former German president Theodor Heuss' statement that "Quality is what is decent". But here we are talking about a standard, and we can be certain that, for the concept of quality, a standard certainly exists. And we will not be disappointed in this respect either by ISO 8402, where the definition of quality used is the "totality of characteristics of an entity that bear on its ability to satisfy stated and implied needs".

In order to understand the standards series, we also need to understand the terms *quality planning, quality inspection* and *quality control (QC)*, which are subsumed under the generic term *quality management*. In quality planning, inputs are defined. Internal inputs include specifications, while external inputs might include the observance of statutory regulations or the definition of customer wishes.

In the case of quality inspection, we must determine to what extent an entity meets the quality requirements. QC, on the other hand, contains the measures to be introduced if test results cause us to intervene by way of prevention, monitoring or correction in order to ensure that the quality requirements are met. ISO 9000 now provides indications as to how such a quality system is to be defined.

In order to understand the standard, it is necessary to know how ISO 9000 came about and what predecessors ISO 9000 has had. ISO stands for "International Standardization Organization" and constitutes the worldwide standardization body. Basically, one must admit that quality systems cannot be standardized; however, generally applicable recommendations and guidelines for quality systems and their individual elements *can* be standardized. The first efforts toward uniform bodies of regulations for quality assurance (QA) were made in the fifties. The objective then was not to replace the product-specific quality standards, but to supplement them with a body of regulations that precisely defined the framework for the consistent implementation of the product-specific standard in the enterprise. The QA works first appeared in the United States, from there they came to Europe. The foundation was laid in the 1950s, first by the military sector and later by the nuclear power industry, both of which had very stringent safety requirements. The Allied Quality Assurance Publications (AQAP) of North Atlantic Treaty Organization (NATO) can be considered a predecessor of today's ISO 9000. In AQAP, uniform inputs were defined for the NATO countries with reference to quality assurance of the services in the case of military equipment.

An additional body of standards in this context is CAN 3-Z 299.1 to 3-Z 299.4. These standards were initially designed for nuclear power station construction and later used also for other production sectors. Similar guidelines also existed in Germany, one of which was KTA 1401, published by the Kerntechnischer Ausschuß (nuclear engineering committee). These KTA rules were restricted exclusively to components and systems for nuclear power stations which were important from the point of view of safety. Also, classical quality assurance guidelines appeared in other sectors with fewer safety requirements. Q 101 is a typical company-specific quality assurance guideline, declared binding by the Ford company in 1990. These developments have one thing in common: over the course of time, one came to realize that a uniform regulation can only be of advantage to all concerned. As a result, ISO 9000 appeared in the 1980s (see Figure 3.1). This was to define a worldwide, uniform, intersectoral standard for quality assurance. Table 3.1 illustrates that ISO 9000 has succeeded in defining a worldwide standard.

In one of his books, Scott Adams (1997) formulated my favorite hypothesis regarding how ISO 9000 came about (Figure 3.2).

THE OBJECTIVES OF ISO 9000

Everybody knows at least one company whose product quality has either experienced no change or deteriorated since introducing ISO 9000. The German foreword to ISO 9000-1 says that the fulfillment of the requirement with respect to QA in the standards ISO 9001, ISO 9002, and ISO 9003 is designed to build trust in the organization's capability. Furthermore, it says that the observance of the standards of quality management is therefore no direct proof that the product offered meets the quality requirement. Already in this German foreword, ISO 9000 says that not the quality of the product, but rather the quality of the system, is the focus of attention. This statement is a very important aspect of ISO 9000, but unfortunately one which is often neglected. Later on we will see that a combination of GMP, i.e., product and process quality, with ISO 9000, that is, system quality, is possible. However, this requires absolute observance of the two differing approaches! The mistake often made is to overlook the differing approaches of these two bodies of regulations.

We often hear it said that an enterprise has established a quality system according to ISO 9000. The German foreword of ISO 9000-1 provides a clear answer here. The international standards of the ISO 9000 series describe which elements quality systems should contain, but not how a specific organization achieves these elements. You may protest quite rightly that in real life one finds manuals structured according to the 20 chapters of ISO 9001. This is some sort of standardization of quality systems, or at least of their manuals. This circumstance is, however, essentially due to a misinterpretation of ISO 9000.

It may of course make sense to divide the manual and thus the essential elements of a quality system into 20 chapters. This is how many international groups of companies proceed, in order to achieve uniform structures in their various business units and at their various locations.

In many cases, however, it may make more sense to choose a completely different structure. If we look at the development of QA systems containing ISO 9000 elements, we will see that more and more of these systems are structured with a product and process

Figure 3.1. ISO 9000 framework of the standardization organization.

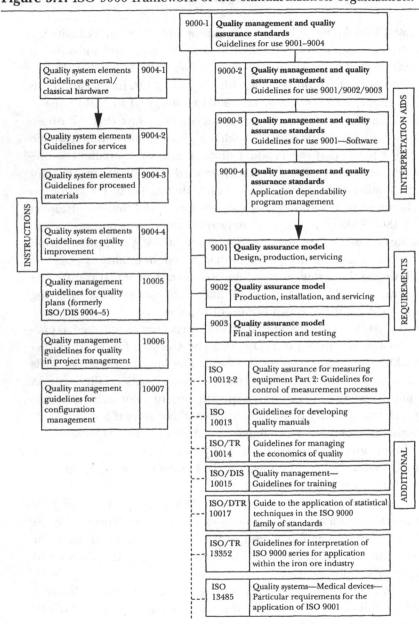

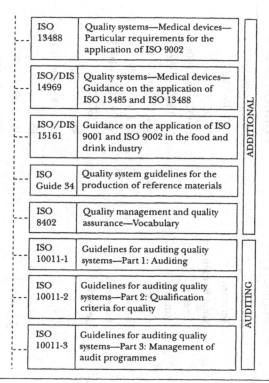

ISO 13488	Quality systems—Medical devices—Particular requirements for the application of ISO 9002	
ISO/DIS 14969	Quality systems—Medical devices—Guidance on the application of ISO 13485 and ISO 13488	ADDITIONAL
ISO/DIS 15161	Guidance on the application of ISO 9001 and ISO 9002 in the food and drink industry	
ISO Guide 34	Quality system guidelines for the production of reference materials	
ISO 8402	Quality management and quality assurance—Vocabulary	
ISO 10011-1	Guidelines for auditing quality systems—Part 1: Auditing	
ISO 10011-2	Guidelines for auditing quality systems—Part 2: Qualification criteria for quality	AUDITING
ISO 10011-3	Guidelines for auditing quality systems—Part 3: Management of audit programmes	

ISO as a standardization organization processes a whole series of standardization projects which in turn are processed by individual "Technical Committees". The ISO Technical Committee TC 176 was and is responsible for ISO 9000 within the framework of the standardization organization. In order to avoid defining a completely new and unknown standard, the standards known at the time (used mainly in the military and nuclear power sectors) were used as the foundation for ISO 9000.

orientation. This is also quite understandable and useful since, after all, an enterprise in the automobile sector requires a very different process and therefore a completely different system structure than an enterprise producing pharmaceutical products. Summarizing the above, ISO 9000 can bring together the sum of the factors which can influence system quality. This means that ISO 9000 is not a description of a system; it merely provides the totality of the elements which should be observed in a quality system.

Table 3.1. Quality system standards of member states.

Member states ISO CEN*/CENELEC** COPANT***	Quality system standards ISO 9000 EN 29000 COPANT-ISO 9000	Member states ISO CEN*/CENELEC** COPANT***	Quality system standards ISO 9000 EN 29000 COPANT-ISO 9000
Australia	AS 3900	France	NF-EN 29000
Austria	Ö Norm EN 29000	Germany	DIN ISO 9000
Barbados	BNS180:1992	Great Britain	BS 5750:1987:Pt 0
Belgium	NBN-EN 29000	Greece	ELOT EN 29000
Brazil	NB 9000:1990	Hungary	MI 18990-1988
Canada	Q 9000	Iceland	IST ISO 9000:1987
Colombia	ICONTEC-ISO 9000	India	IS 14000:1988
Chile	NCH-ISO 9000	Indonesia	SNI 19-9000-1991
China	GB/T10300.1-88	Ireland	IS/SO 9000
Cuba	NC-ISO 9000	Israel	SI 2000:1990
Cyprus	CYS ISO 9000	Italy	UNI/EN 29000-1987
Czech Republic	CNS ISO 9000	Jamaica	JS
Denmark	DS/ISO 9000	Japan	JIS Z 9000-1991
Egypt	ES/ISO 9000	Korea, Rep. of	KS A 9000-1992
Finland	SFS-ISO 9000	Malaysia	MS-ISO 9000-1991

Member states	Quality system standards	Member states	Quality system standards
Mexico	NOM-CC-2	Sri Lanka	SLS 825:Part 2:1988
Netherlands	NEN-ISO 9000	Sweden	SS-ISO 9000:1989
New Zealand	NZS 9000:1990	Switzerland	SN EN 29000:1990
Norway	NS-ISO 9000:1988	Tanzania	TZS 500:1990
Pakistan	PS:3000:90	Thailand	TISI ISO 9000
Philippines	PNS ISO 9000:1989	Trinidad/Tobago	TTS 165 400:1988
Poland	ISO 9000	Tunisia	NT 110.18-1987
Portugal	NP EN 29000	Turkey	TS-ISO 9000
Romania	RS ISO 9000	United States	ANSI/ASQC Q90
Russia		Uruguay	UNIT-ISO 9000-91
Singapore	SS/ISO 9000:1988	Venezuela	COVENIN-ISO 9000
Slovakia	CNS ISO 9000	Yugoslavia	JUS-ISO 9000
South Africa	SABS/ISO 9000	Zimbabwe	SAZS 300:1990:P5
Spain	UNE 66 900		

*European Committee for Standardization
**European Committee for Electrotechnical Standardization
***Pan American Standards Commission

Figure 3.2. Scott Adams on ISO 9000.

My theory is that a bunch of Europeans were bored, and after drinking a few too many Heineken beers, they decided to play a mean trick on all the world's large companies. The trick was ISO 9000. It was called that because of the number of beers they drank that night ("ISO" is either an incomprehensible combination of letters or possibly one of 400 European slang expressions for "Is that my beer?").

THE CONTENT OF ISO 9000-1

The table of contents of ISO 9000-1 is shown in Table 3.2.

As one can see from the table of contents, the objective of this standard is to explain the principal quality management concepts, their contexts, differences and interrelationships within the ISO 9000 series. This is to provide a guideline for the selection and use of the standards. A complete commentary on this standard would doubtless be boring and of little help to you, the user. For this reason, in the following I would like to interpret only the really important statements. Section 4.1 lists the key quality-related objectives and the responsibilities which an organization should aim for (Table 3.3).

Although I am no friend of extensive definitions, Section 4.1 explains the objectives of ISO 9000 very well. Sections 4.6 to 4.8 describe the fundamental concept of the ISO 9000 series. We read there, among other things, that every organization seeking to establish a quality system needs to identify, organize and manage its network of processes and interfaces. The processes and its interfaces should therefore be subjected to analysis and continuous improvements.

Table 3.2. The table of contents of ISO 9000-1.

0	Introduction
1	Scope
2	Normative reference
3	Definitions
4	Principal concepts
4.1	Key objectives and responsibilities for quality
4.2	Stakeholders and their expectations
4.3	Distinguishing between quality system requirements and product requirements
4.4	Generic product categories
4.5	Facets of quality
4.6	Concept of a process
4.7	Network of processes in an organization
4.8	Quality system in relation to the network of processes
4.9	Evaluating quality systems
5	Roles of documentation
5.1	Value of documentation
5.2	Documentation and evaluation of quality systems
5.3	Documentation as a support for quality improvement
5.4	Documentation and training
6	Quality system situations
7	Selection and use of International Standards on quality
7.1	General
7.2	Selection and use
7.3	Application guidelines
7.4	Software
7.5	Dependability
7.6	Quality assurance: design, development, production, installation, and servicing
7.7	Quality assurance: production, installation, and servicing
7.8	Quality assurance: final inspection and test

Continued on the next page.

The next sections explain how quality systems are evaluated by means of audits, what functions documentation has in a quality system, what measures can be introduced for quality improvement and how training courses are to be conducted and documented. ISO 9000-1 provides clear statements only under Section 7, "Selection and use of International Standards on quality". These items are suitable as a reference work. A total of four annexes are intended to provide the reader with additional information.

From my own point of view, Annex D, an informative annex with no normative character, is a real help. The chart from Annex D shown in Table 3.3 actually provides what ISO 9000-1 promises. It shows what requirements with respect to demonstration of con-

Table 3.3. Section 4.1 of ISO 9000.

4   Principal concepts

4.1. Key objectives and responsibilities for quality

An organization should:

   a. achieve, maintain and seek to improve continuously the quality of its products in relationship to the requirements for quality;

   b. improve the quality of its own operations, so as to meet continually all customers' and other stakeholders' stated and implied needs;

   c. provide confidence to its internal management and other employees that the requirements for quality are being fulfilled and maintained, and that quality improvement is taking place;

   d. provide confidence to the customers and other stakeholders that the requirements for quality are being, or will be, achieved in the delivered product;

   e. provide confidence that quality system requirements are fulfilled.

formance ISO 9001, 9002, and 9003 contain and in what points they differ. Furthermore, reference is made to the quality guidelines of ISO 9004-1.

The Models for Demonstration of Conformity: ISO 9001, 9002 and 9003

ISO 9001, 9002, and 9003 can be defined below:

- ISO 9001 quality systems—model for QA in design, development, production, installation and servicing.

- ISO 9002 quality systems—model for QA in production, installation and servicing.

- ISO 9003 quality systems—model for QA in final inspection and testing.

The aforementioned models are identical in structure. The difference is that some of the elements contained in ISO 9001 are not covered by ISO 9002 and ISO 9003. Table 3.4 shows that ISO 9001

Table 3.4. Cross-reference list of clause numbers for corresponding topics.

| Requirements | | | Application guide | | | |
ISO 9001	ISO 9002	ISO 9003	ISO 9000-2	Clause title in ISO9001	QM guidance	Road map
4.1	●	◉	4.1	Management responsibility	4	4.1; 4.2; 4.3
4.2	●	◉	4.2	Quality system	5	4.4; 4.5; 4.8
4.3	●	◉	4.3	Contract review	○	8
4.4	○	○	4.4	Design control	8	
4.5	●	●	4.5	Document and data control	5.3; 11.5	
4.6	●	○	4.6	Purchasing	9	
4.7	●	●	4.7	Control of customer-supplied product	○	
4.8	●	◉	4.8	Product identification and traceability	11.2	5
4.9	●	○	4.9	Process control	10;11	4.6; 4.7
4.10	●	◉	4.10	Inspection and testing	2	
4.11	●	●	4.11	Control of inspection, measuring, and test equipment	13	
4.12	●	●	4.12	Inspection and test status	11.7	
4.13 ●●	◉	4.13		Control of nonconforming product	14	

Requirements			Application guide	Clause title in ISO9001	QM guidance	Road map
ISO 9001	ISO 9002	ISO 9003	ISO 9000-2			
4.14	●	●	◉	4.14	Corrective and preventive action	15	
4.15	●	●	●	4.15	Handling, storage, packaging, preservation and delivery	10.4;16.1;16.2	
4.16	●	●	◉	4.16	Control of quality records	5.3; 17.2; 17.3	4.9
4.17	●	●	◉	4.17	Internal quality audits	5.4	5.4
4.18	●	●	◉	4.18	Training	18.1	
4.19	●	●	O	4.19	Servicing	16.4	
4.20	●	●	◉	4.20	Statistical techniques	20	
					Quality economics	6	
					Product safety	19	
					Marketing	7	

Key:

● = Comprehensive requirement

◉ = Less-comprehensive requirement than ISO 9001 and ISO 9002

O = Element not present

is the most extensive model and covers all 20 elements. This model is most suitable for enterprises seeking to demonstrate the conformance of a complete process, ranging from research and development to production and servicing. ISO 9002 has an identical scope of regulation, but does not cover the research and development sector. This means that ISO 9002 is the model of choice for contract manufacturers. ISO 9003 occurs very rarely in practice, as the regulations refer only to final inspection and testing. In the following, therefore, the most comprehensive model, ISO 9001, will be described with all its elements. And to say it again: certification can only take place according to ISO 9001, 9002, or 9003. ISO 9000 and ISO 9004 are not models for demonstrating conformance, but rather serve only for the establishment of quality systems. Table 3.5 lists the 20 elements described in ISO 9001.

When you read all the way through the ISO 9001 standard, it is difficult to make the connection to pharmaceutical-specific processes. One reason for this is that the selected terms are designed so that they can be used intersectorally, across a variety of fields. I will therefore describe each individual element briefly, including the objective, the actual content-related requirements and the corresponding implementations as we know them in the pharmaceutical industry, in particular those of the EC Guide to GMP. This will give you a cross-reference from each element of ISO 9001 to the GMP Guidelines and their implementation in practice.

A word about my interpretation: Unfortunately, there are a large number of possible interpretations of ISO 9000 in practice. There is no generally recognized written commentary. To give a drastic example: Some ISO 9000 auditors are of the opinion that ISO 9001 certification is not possible for a pharmaceutical contract manufacturer and refuse to issue an ISO 9001 certification since, according to their interpretation, only ISO 9002 certification is possible. Other auditors argue that, even with very little galenical development, Element 4.4, "Design control", is covered, thus making certification possible.

Table 3.5. Elements of ISO 9001.

0	Introduction
1	Scope
2	Normative reference
3	Definitions
4	Quality system requirements
4.1	Management responsibility
4.2	Quality system
4.3	Contract review
4.4	Design control
4.5	Document and data control
4.6	Purchasing
4.7	Control of customer-supplied product
4.8	Product identification and traceability
4.9	Process control
4.10	Inspection and testing
4.11	Control of inspection, measuring and test equipment
4.12	Inspection and test status
4.13	Control of nonconforming product
4.14	Corrective and preventive action
4.15	Handling, storage, packaging, preservation and delivery
4.16	Control of quality records
4.17	Internal quality audits
4.18	Training
4.19	Servicing
4.20	Statistical techniques
Annex A	Bibliography

Nevertheless, the accreditation agencies are trying to define a uniform policy. But there are plenty of different interpretations, as has been explained above. In this respect, I describe my experience as that of a "Qualified ISO 9000 Auditor" who has undergone personal certification according to ISO 10 011 Part 2. My interpretations are based on a number of pharmaceutical enterprises which I have accompanied in ISO 9000 projects.

ELEMENT I: MANAGEMENT RESPONSIBILITY

Objective: To define responsibility for quality in management.

DIN ISO 9001		EC Guide to GMP
4.1	Management responsibility	——————
4.1.1	Quality policy	·l.
4.1.2.	Organization	
4.1.2.1	Responsibility and authority	Chapter 1 and 2 principles 2.2–2.7
4.1.2.2	Resources	2.1
4.1.2.3	Management representative	
4.1.3	Management review	·l.

Legend: —— = no requirements defined, only headlines; ·l. = no similar requirement with the EC GMP Guide; the numbers given refer to the relevant requirement of the EC GMP Guide.

Content-Related Requirement

The content of this element is the definition of quality policy. In addition, the management must equip a corresponding organization with staff and resources in order to implement the quality measures. Also, a management representative is required. Another of the core elements is the evaluation of the quality system by the company management, for which a specific procedure needs to be defined.

Overlap with the EC Guide to GMP

The differing objectives of the ISO 9000 series and the EC Guide to GMP are most conspicuous in this element. Of course, the EC Guide to GMP requires that sufficient and qualified personnel be present in the individual departments. And the corresponding appointment of "qualified persons" is also extensively regulated. It does not contain the requirement of a quality policy laid down in writing, an appointment of a quality representative, or the requirement of a formalized evaluation method for review by the management. As I explained in my introduction, the reason for this lies in the various quality approaches (product/process quality approach versus system quality approach).

ELEMENT II: QUALITY SYSTEM

Objective: To establish and maintain a
documented quality system.

	DIN ISO 9001	EC Guide to GMP
4.2	Quality system	——————
4.2.1	General	1.1/1.2
4.2.2	Quality system procedures	4.26
4.2.3	Quality planning	5.21–5.24/5.37

Legend: —— = no requirements defined, only headlines; ·l· = no similar requirement with the EC GMP Guide; the numbers given refer to the relevant requirement of the EC GMP Guide.

Content-Related Requirement

According to this element, the enterprise must exhibit a quality manual and quality system procedures.

A further section deals with the topic of quality for planning, which must define and document how the enterprise fulfils quality requirements in the individual products. These quality plans can also be shown in the form of references to relevant documented procedures. The element specifies that measures must be taken as early as possible to ensure that the production process, the inspection and testing techniques, all other relevant measures, and the relevant documentations and quality records are suitable for fulfilling the quality requirements of the product.

Overlap with the EC Guide to GMP

The basic requirements for quality systems are found in Chapter 1 of the EC Guide to GMP. The corresponding product-related documents are explained in Chapter 4. In analogy to what was said under "Management responsibility" in Element 1, here too the focus of ISO 9000 is on the system elements. In this case, the specific focus is on the necessity of a quality manual and quality system procedures. The latter is also found in Chapter 4 of the EC Guide to GMP.

Chapter 3, "Premises and Equipment", and Chapter 5, "Production", contain more extensive regulations for the field of quality planning. Essentially, one can subsume the activities of the scale-up process, process validation, qualification of plant and equipment, under quality planning. The regulations on this in the EC Guide to GMP are more detailed than those in the ISO 9000 series, which does not acknowledge these specific requirements.

ELEMENT III: CONTRACT REVIEW

Objective: Only ensured quality may be promised to the customer in contracts.

	DIN ISO 9001	EC Guide to GMP
4.3	Contract review	————
4.3.1	General	7.1–7.2
4.3.2	Review	7.3–7.5
4.3.3	Amendment to a contract	·l.
4.3.4	Records	7.10–7.12

Legend: ———— = no requirements defined, only headlines; ·l. = no similar requirement with the EC GMP Guide; the numbers given refer to the relevant requirement of the EC GMP Guide.

Content-Related Requirement

The "Contract Review" element contains a detailed description of the review of a contract and the mode of procedure for amendments to a contract, and also defines how records must be made on contract reviews.

Overlap with the EC Guide to GMP

Chapter 7 of the EC Guide to GMP, "Contract Manufacture and Analysis", contains requirements concerning a written contract. Statements on content and formulation are also made in the following. In this element too, the differing objectives of the EC Guide to GMP and the ISO 9000 standards series become clear. While Chapter 7 of the EC Guide to GMP also specifies in detail (Section 7.12) who is responsible for the purchase of materials, the inspection and release of materials, the implementation of production and quality controls, and so on. ISO 9000 regulates general contractual matters in the section "Contract review". In Chapter 7 of the EC Guide to GMP, for instance, one finds no detailed regulations as to how a contract review must take place. The most important difference, however, is that there are several departments within an enterprise that are concerned with the topic of contract review. This means, in addition to manufacture and contract manufacture review, contract review in purchasing is to be assigned just like contract review in sales. The EC Guide to GMP and other pharmaceutical bodies of regulations offer no further statements on this.

ELEMENT IV: DESIGN CONTROL

Objective: Systematic design and development planning.

DIN ISO 9001		EC Guide to GMP
4.4	Design control	
4.4.1	General	·l.

Continued on the next page.

Continued from the previous page.

DIN ISO 9001		EC Guide to GMP
4.4.2	Design and development planning	·l.
4.4.3	Organisational and technical interfaces	·l.
4.4.4	Design input	·l.
4.4.5	Design output	·l.
4.4.6	Design review	·l.
4.4.7	Design verification	·l.
4.4.8	Design validation	·l.
4.4.9	Design changes	·l.

Legend: —— = no requirements defined, only headlines; ·l. = no similar requirement with the EC GMP Guide; the numbers given refer to the relevant requirement of the EC GMP Guide.

Content-Related Requirement

Element IV is one of the most extensive system elements of ISO 9000. It describes the complete process of the planned and documented research and development (R&D), ranging from planning, definition of inputs, documentation of the results and inspection and certification of the design, to design validation, which defines the transition into the field of process control. Finally the handling of design changes, including their identification, documentation and review, is defined.

Overlap with the EC Guide to GMP

Without a doubt, the assignment of content to this element is altogether the most difficult. The reason for this is that all the requirements of Element IV lie outside the scope of regulation of the EC Guide to GMP. Only Annex 14 to the EC Guide to GMP "Manufacture of Investigational Medicinal Products" covers a small area of Element IV of ISO 9001. Nevertheless, we find in other pharmaceutical guidelines a large quantity of information that can be subsumed under Element IV. This includes *inter alia* the conducting of clinical trials, the individual steps within the framework of the registration procedure, the defined processes within the framework of a scale-up process and ranges to process validation.

In many cases this process is not structured and documented to the extent necessary within the scope of ISO 9000, owing to the fact that, in addition to the classical production sector, above all the Research & Development (R&D) sector is involved. In many cases, therefore, although one can find documentation on Element 4.4.2 (design and development planning), these process documents are rarely integrated in a company-wide quality system. Element IV can therefore provide interesting suggestions for pharmaceutical enterprises for bundling in a uniform system the large quantity of documentation compiled on the basis of pharmaceutical guidelines.

ELEMENT V: DOCUMENT AND DATA CONTROL

Objective: Valid documents are to be provided by authorized personnel.

	DIN ISO 9001	EC Guide to GMP
4.5	Document and data control	————
4.5.1	General	chapter 4 especially 4.4
4.5.2	Document and data approval and issue	4.2/4.3
4.5.3	Document and data changes	4.5/5.15

Legend: ——— = no requirements defined, only headlines; ·l. = no similar requirement with the EC GMP Guide; the numbers given refer to the relevant requirement of the EC GMP Guide.

Content-Related Requirement

Like Element XI, "Control of inspection, measuring and test equipment", Element V "Document and data control" is often known as the auditors' playground. The reason for this is that the standard requires in detail the review, approval, issue and revision control of all documents and data. Element V serves as an indicator for the effective functioning of a quality system.

Overlap with the EC Guide to GMP

Chapter 4 of the EC Guide to GMP contains similar requirements. The focus in the EC Guide to GMP is on product-related documents, that is, specifications, manufacturing formulae and instructions and records. There are only a few system documents; procedures are specifically mentioned in 4.19–4.26. ISO 9001 considers essentially the system documents here, that is, the manual and documented procedures which are organized in a product-independent manner.

ELEMENT VI: PURCHASING

Objective: To ensure the quality of purchased product.

DIN ISO 9001		EC Guide to GMP
4.6	Purchasing	
4.6.1	General	5.25/5.40
4.6.2	Evaluation of subcontractors	5.26 Supplementary Guideline No. 8, Item 3
4.6.3	Purchasing data	4.10–4.13
4.6.4	Verification of purchased goods	——
4.6.4.1	Supplier verification at subcontractor's premises	7.12
4.6.4.2	Customer verification of subcontracted product	·l.

Legend: —— = no requirements defined, only headlines; ·l. = no similar requirement with the EC GMP Guide; the numbers given refer to the relevant requirement of the EC GMP Guide.

Content-Related Requirement

A qualified purchasing process, according to ISO 9000, requires that a systematic evaluation of suppliers take place. The purchasing data, which must be given to the supplier as a clear specification, is also an important aspect. The item "Verification of purchased product" deals with two special cases. One is the special case wherein the manufacturer performs the verification directly at the supplier's premises, and the other occurs when the manufacturer's customer performs the verification directly at the supplier's premises.

Overlap with the EC Guide to GMP

Chapter 6, "Quality Control", contains detailed statements about specifications. The procedure for the evaluation of suppliers (supplier qualification) is also described in the EC Guide to GMP. There is a difference in the way the purchasing sector is viewed. While the EC Guide to GMP defines these statements almost exclusively from the point of view of QC, owing to the special responsibility of the Control Manager, ISO 9000 describes the element from the point of view of the purchasing department.

ELEMENT VII: CONTROL OF CUSTOMER-SUPPLIED PRODUCT

Objective: To ensure the quality of the supplied product.

	DIN ISO 9001	EC Guide to GMP
4.7	Control of customer-supplied product	·l.

Legend: —— = no requirements defined, only headlines; ·l. = no similar requirement with the EC GMP Guide; the numbers given refer to the relevant requirement of the EC GMP Guide.

Content-Related Requirement

Element VII describes a process which occurs rather seldom in practice, almost exclusively in the case of contract manufacturers. Element VII specifies that the starting materials, packaging materials, and so on, supplied by a customer for production should also be systematically verified, stored and controlled.

Overlap with the EC Guide to GMP

As already mentioned, this special case is generally found in the case of contract manufacturers. The contract giver supplies the contract acceptor (contract manufacturer) with certain starting materials. As a rule, this activity is defined by a contract delimiting responsibility that must be drawn up according to Chapter 7 of the EC Guide to GMP. In this case the contract manufacturer often proceeds with an identity control only as verification of the purchased goods.

ELEMENT VIII: PRODUCT IDENTIFICATION AND TRACEABILITY

Objective: To avoid mixups and to ensure the traceability of products.

DIN ISO 9001		EC Guide to GMP
4.8	Product identification and traceability	4.21/4.25/5.42/7.13/ 8.1–8.15

Legend: —— = no requirements defined, only headlines; ·l. = no similar requirement with the EC GMP Guide; the numbers given refer to the relevant requirement of the EC GMP Guide.

Content-Related Requirement

The products must be able to be clearly classified in all phases of production and delivery. This requires the clear identification and recording for traceability.

Overlap with the EC Guide to GMP

In their content-related requirements, the regulations in Chapter 5, "Production" and Chapter 8, "Complaints and Product Recall" far exceed those defined in ISO 9000 for the identification of the status of a product. The complete identification of the individual starting materials and the absolutely necessary records about the delivered products have been dealt with in detail in the EC Guide to GMP

ELEMENT IX: PROCESS CONTROL

Objective: Production processes must be planned and controlled with respect to quality.

DIN ISO 9001		EC Guide to GMP
4.9	Process control	3.6–3.16/4.1/3.34–3.44/4.14–4.16/ 4.26/4.27/5.1/5.5/5.6/5.9/5.11/ 5.14–5.20/5.32/5.34–5.35

Legend: —— = no requirements defined, only headlines; ·l. = no similar requirement with the EC GMP Guide; the numbers given refer to the relevant requirement of the EC GMP Guide.

Content-Related Requirement

Element IX of ISO 9001 is kept relatively brief. The essence of the requirements is that production must be planned, monitored, and controlled. For this, relevant quality plans and documented procedures must be compiled. Where appropriate, processes and equipment must also be approved. The topic of maintenance of equipment is also located in Element IX.

Overlap with the EC Guide to GMP

Chapter 9, "Process Control", makes clear again how the objectives of the EC Guide to GMP and those of ISO 9001 differ. The EC Guide to GMP contains a large number of regulations, in Chapter 3, "Premises and Equipment", Chapter 4, "Documentation", and Chapter 5, "Production", which far exceed those defined in Element IX of ISO 9001. One can say that the regulation on the topic of process control supplies no new findings for a pharmaceutical QA system.

ELEMENT X: INSPECTION AND TESTING

Objective: To prove the fulfillment of specified requirements on the basis of documented inspection and testing procedures.

DIN ISO 9001		EC Guide to GMP
4.10	Inspection and testing	———
4.10.1	General	1.4/6.1–6.4/6.5–6.6/ 3.26–3.29
4.10.2	Receiving inspection and testing	·l.
4.10.2.1	Quarantine status for deliveries not yet inspected and tested	5.5/5.30–5.31 4.22/(3.22)/4.19/4.20/ 5.3/5.27–5.28
4.10.2.2	Reduced/restricted receiving of inspection and testing	8. Supplementary Guideline Item 2
4.10.2.3	Risk withdrawal	·l.
4.10.3	In-process inspection and testing	3.17/4.23/5.38/5.39/ 6.11–6.14/6.18
4.10.4	Final inspection and testing	4.23–4.24/5.58
4.10.5	Inspection and test records	6.7–6.10/6.17

Legend: ——— = no requirements defined, only headlines; ·l. = no similar requirement with the EC GMP Guide; the numbers given refer to the relevant requirement of the EC GMP Guide.

Content-Related Requirement

Element X, "Inspection and Testing", distinguishes between receiving inspection and testing, in-process inspection and testing, final inspection and testing, and inspection and test records. The receiving inspection and testing must be carried out in accordance with a documented procedure, and it must be ensured that the products delivered are not used before release. Section 4.10.2.2 defines that the amount of inspection shall be based on how the supplier itself inspects and what proofs it supplies of the conformity of the product. ISO 9000 requires that in-process inspection and testing take place in accordance with appropriate written instructions.

The same applies to the final inspection and testing, in which it is also required that a product may not be dispatched before inspection and testing has been completed. The item "Inspection and test records" defines once again that relevant proofs must be kept in order to document the inspection and testing in a comprehensible manner. The inspection and test records must name the responsible inspector.

When deviations occur it must be ensured that one proceeds in accordance with Element XIII, "Control of nonconforming product".

Overlap with the EC Guide to GMP

At first glance it appears that the entire scope of regulation detailed in Element X, "Inspection and testing", is also covered by the EC Guide to GMP. Upon closer inspection, however, we see that the regulations contained in Chapter 6 and further-reaching regulations in Chapters 3 and 4 do not fully cover the regulation of ISO 9001. The item 4.10.2.2, concerning reduced and restricted receiving inspection and testing, is not found in the EC Guide to GMP. Only Item 2 of Supplementary Guideline No. 8 contains regulations to this effect. The sub-item 4.10.2.3 (risk withdrawal) finds no mention in the EC Guide to GMP. Nevertheless, risk withdrawl is a common practice in Europe. But there is an FDA statement that follows.

> It *is not acceptable* to use drug components, containers, or closures prior to completion of all testing to determine conformance to established specifications. The preamble to 21 CFR 211.84, clearly states that the use of drug components, containers, or closures prior to completion of testing for conformity to specifications *violates the precept of good quality control* because *untested and possibly noncomplying materials would be used in drug product processing.* This type of procedure substantially increases the risk to the consumer that an unsatisfactory lot might erroneously be released (CDER-Human Drug cGMP-Notes, 1996).

On the whole, it can be said that the detailed regulation of the EC Guide to GMP far exceeds that of ISO 9001. Nevertheless, the individual regulations, as the above references show, are scattered throughout the EC Guide to GMP.

ELEMENT XI: CONTROL OF INSPECTION, MEASURING AND TEST EQUIPMENT

Objective: To control the suitability of all inspection, measuring, and test equipment, including software which influences product quality.

DIN ISO 9001		EC Guide to GMP
4.11	Control of inspection, measuring, and test equipment	————
4.11.1	General	3.40
4.11.2	Control procedure	6.19–6.21/3.41

Legend: ———— = no requirements defined, only headlines; ·l. = no similar requirement with the EC GMP Guide; the numbers given refer to the relevant requirement of the EC GMP Guide.

Content-Related Requirement

Element XI, which is also often termed the auditors' playground, stipulates in great detail that all inspection, measuring, and test equipment be recorded, that relevant identifications and calibration procedures exist, and that calibration be conducted with certificated resources. Regulations have also been made with respect to the storage of records. Furthermore, reference is made to ISO 10 012 as a guideline for the calibration of inspection, measuring and test equipment.

Overlap with the EC Guide to GMP

In the EC Guide to GMP, the topic of control of inspection, measuring and test equipment in the narrower sense is only mentioned under items 3.40 and 3.41. The description of the requirement for control of inspection, measuring and test equipment in ISO 9001 greatly exceeds that which the EC Guide to GMP describes. Items such as establishment of calibration intervals, relationship to international and national standards, definition of calibration processes, identification of inspection, measuring and test equipment, calibration records, regulations concerning environmental conditions of calibrations, storage and handling of inspection, measuring and test equipment are all missing in the EC Guide to GMP.

Nor does the EC Guide to GMP contain requirements for evaluating and documenting the validity of results of previous inspection and test results if inspection, measuring or test equipment is found to be out of calibration. On the whole, one can say that Element XI supplies a very good guideline for the calibration of measuring equipment and that ISO 10 012 must by all means be recommended for performing calibrations.

ELEMENT XII: INSPECTION AND TEST STATUS

Objective: To identify of the inspection and test status of the product.

DIN ISO 9001		EC Guide to GMP
4.12	Inspection and test status	4.24/5.12–5.13/5.58

Legend: —— = no requirements defined, only headlines; ·l. = no similar requirement with the EC GMP Guide; the numbers given refer to the relevant requirement of the EC GMP Guide.

Content-Related Requirement

The inspection and test status of a product must be identified so that the inspection and test status can be ascertained at any time. Dispatch and use may only take place after release. The topic of release under an authorized concession is also mentioned.

Overlap with the EC Guide to GMP

The topic of inspection and test status is explained in Chapter 4.24 of the EC Guide to GMP. Furthermore, one finds regulations in Chapter 5, "Production", under Sections 5.12 and 5.13 which are almost identical to the requirements of ISO 9001. There are therefore no new regulations in ISO 9001 which reach further than the EC Guide to GMP.

ELEMENT XIII: CONTROL OF NONCONFORMING PRODUCT

Objective: If nonconforming product is identified, a procedure for its identification, documentation, evaluation and segregation must take place.

DIN ISO 9001		EC Guide to GMP
4.13	Control of noncon-forming product	
4.13.1	General	(3.23)/5.61 (1.3/VI), (1.4/IV), (1.4/VI)
4.13.2	Review and disposit-ion of nonconform-ing product	5.62/5.65

Legend: —— = no requirements defined, only headlines; ·l. = no similar requirement with the EC GMP Guide; the numbers given refer to the relevant requirement of the EC GMP Guide.

Content-Related Requirement

ISO 9001 requires a written procedure which defines the responsibility for the evaluation and further mode of procedure for nonconforming product. It is also regulated that accepted nonconformities be recorded, and that reworked and/or repaired product be reinspected according to corresponding documented procedures.

Overlap with the EC Guide to GMP

The EC Guide to GMP regulates the review and disposition of nonconforming product in Chapter 5, Section 5.62 to 5.65. Section 3.23 of Chapter 3, "Premises and Equipment", also regulates the separate storage of rejected product. This element of ISO 9000 makes additional requirements.

ELEMENT XIV: CORRECTIVE AND PREVENTIVE ACTION

Objective: To eliminate the causes of nonconformity in order to prevent the repetition of the nonconformity.

	DIN ISO 9001	EC Guide to GMP
4.14	Corrective and preventive action	————————
4.14.2	Corrective action	Chapter 8, 5.39/ 5.55–5.56, 5.15
4.14.3	Preventive action	·l.

Legend: ——— = no requirements defined, only headlines; ·l. = no similar requirement with the EC GMP Guide; the numbers given refer to the relevant requirement of the EC GMP Guide.

Content-Related Requirement

The element requires suitable corrective and preventive action. Therefore the causes of product nonconformities must be ascertained, and actions recorded and implemented in accordance with the risk. In addition, the efficacy of the action must be controlled. The changes resulting from corrective and preventive action must be implemented and the documented procedures recorded.

Overlap with the EC Guide to GMP

The review and recording of nonconformities are found, for instance, in Chapter 5.39 with reference to the recording and investigation of yield results. Section 5.55 also requires that, after unusual events, action must be conducted and recorded. Section 5.56 requires that "any significant or unusual discrepancy observed during reconciliation of the amount of bulk product and printed packaging materials and the number of units produced should be investigated and satisfactorily accounted for before release". Many regulations of Chapter 8, "Complaints and Product Recall", can also be subsumed under the aspect of corrective action. On the topic of preventive action, ISO 9001 requires a systematic analysis of possible preventive action. Regulations are, of course, found in Chapter 9, "Self-inspection", validation in Chapter 5, Section

5.21 to 5.24, and the topic of in-process and environmental controls in Section 5.38, but the requirement of ISO 9001 reaches further here than in the EC Guide to GMP. The reason for this is that a systematic procedure for ascertaining preventive action must be maintained. In this respect the element "preventive action" is to be regarded in addition to the measures of the internal audits of quality planning, and so on.

ELEMENT XV: HANDLING, STORAGE, PACKAGING, PRESERVATION AND DELIVERY

Objective: To avoid damage and deterioration of the quality of the product.

	DIN ISO 9001	EC Guide to GMP
4.15	Handling, storage, packaging, preservation and delivery	——————
4.15.1	General	3.1–3.5/3.30/3.33
4.15.2	Handling	3.6–3.16
4.15.3	Storage	3.18–3.25/4.21/5.7/5.8/ 5.29/5.36/5.41/5.60
4.15.4	Packaging	(4.16)/5.40–5.43/ 5.44–5.57
4.15.5	Preservation	3.19/5.60
4.15.6	Delivery	3.20/4.25

Legend: —— = no requirements defined, only headlines; ·l. = no similar requirement with the EC GMP Guide; the numbers given refer to the relevant requirement of the EC GMP Guide.

Content-Related Requirement

First, it is required that relevant documented procedures for handling, storage, packaging, preservation and delivery exist.

In the case of handling, suitable internal and external logistical channels are to be ensured. Under the heading of "storage", requirements for the condition of the storage rooms are formulated in order to prevent product damage. The sections "Packaging", "Preservation" and "Delivery" require that these processes be controlled and that the product, as long as it is in the sphere of influence of the manufacturer, be suitably stored and deterioration ruled out.

Overlap with the EC Guide to GMP

Regulations on this topic are defined in various forms in the individual chapters of the EC Guide to GMP. As regards delivery, for instance, one finds the regulations in Chapter 3, "Premises and Equipment", under Item 3.20 (requirement for the receiving and dispatch areas). The packaging processes are defined in very great detail in Sections 5.44 to 5.57 of Chapter 5.

ELEMENT XVI: CONTROL OF QUALITY RECORDS

Objective: To ensure proof of the required quality of the product and efficacy of the quality system by means of quality recording.

DIN ISO 9001		EC Guide to GMP
4.16	Control of quality records	3.41/4.6/4.1(records)/ 4.8/4.7/4.9/4.17/4.18/ 4.19/4.20/4.23/4.25/ 4.26/4.28/4.29/5.2/5.33/ 5.55/6.17

Legend: ——— = no requirements defined, only headlines; ·l. = no similar requirement with the EC GMP Guide; the numbers given refer to the relevant requirement of the EC GMP Guide.

Content-Related Requirement

Element XVI requires that quality records be presented for the proof of fulfillment of quality requirements. The term "quality records" also includes the quality records of the supplier. A documented procedure for quality recording must exist which regulates identification, collection, indexing, access, filing, storage, maintenance and disposition.

Overlap with the EC Guide to GMP

Chapter 4, "Documentation", Chapter 5, "Production" and Chapter 6, "Quality Control", mention a large number of quality records. Unlike ISO 9001, the EC Guide to GMP does not define these records as an independent element. If one reads right through the EC Guide to GMP, one finds above all in Sections 4.1, 4.6, 4.7, 4.8 and 4.9 detailed requirements as to how such quality records are compiled, managed and archived.

In addition, in the individual sections of the chapters "Production" and "Quality Control" the EC Guide to GMP defines which individual quality records are to be kept, for instance 4.17, "Batch Processing Records", 4.18, "Batch Packaging Records", 4.19, "Receipt Procedures and Records", 4.20, "Records of Receipt and their Contents", and so forth.

To return to the difference between system quality (ISO 9001) and product/process quality (EC Guide to GMP): Element XVI deals above all with how the administration, collection and archiving of records is to take place (system point of view), while the EC Guide to GMP deals with which quality records are to be made (product/process point of view), and which elements they should contain.

ELEMENT XVII: INTERNAL AUDITS
Objective: To evaluate the current status of the quality
management system.

DIN ISO 9001		EC GMP Guide to GMP
4.17	Internal Audit	Chapter 9

Legend: ⸺ = no requirements defined, only headlines; ·I. = no similar requirement with the EC GMP Guide; the numbers given refer to the relevant requirement of the EC GMP Guide.

Content-Related Requirement

This element defines that internal audits should be conducted in order to evaluate the current status of the quality management system. It is necessary to establish and maintain a documented procedure on how these internal audits should be planned, conducted and documented. The auditor should be independent from those having direct responsibility for the activity being audited. Corrective action and follow-up audits are mentioned as requirements when deficiencies have been evaluated. There are two notes in this element. The first note mentions that the audit reports are an integral part of the input to the management review activities (Element 4.1.3). The second note refers to the ISO 10011 which is a guideline on how quality system audits should be planned and conducted. The ISO 10011 Guideline gives very detailed advice and can be recommended also when ISO compliance is not the objective.

Overlap with the EC Guide to GMP

At first look the requirements of Element 4.17 and Chapter 9 "Self Inspection of the EC GMP Guide" seem to be very similar. But at a second look there several differences. For example: the main focus of ISO 9001 is to determine the effectiveness of the quality system. The EC GMP Guide defines that a detailed evaluation of the current status of the implementation of Good Manufacturing Practices should be the main objective. Personnel, premises and equipment, documentation and production are mentioned again in this chapter in order to stress the importance of the GMP compliance control. But there is not much said about the effectiveness control of the quality system. There is no doubt that these requirements can be combined in order to conduct audits that focus on both GMP and quality system.

Additionally ISO 9000 determines more details on which documents should be established and how to proceed with corrective actions.

ELEMENT XVIII: TRAINING

Objective: To secure and maintain adequate personnel qualification.

DIN ISO 9001		EC Guide to GMP
4.18	Training	2.8/2.9/2.12

Legend: ——— = no requirements defined, only headlines; ·l. = no similar requirement with the EC GMP Guide; the numbers given refer to the relevant requirement of the EC GMP Guide.

Content-Related Requirement

Although ISO 9001 classes training as one of the leading elements, the corresponding requirements are formulated in only three sentences. They require regulations for identifying training needs, implementation of training and records of training.

Overlap with the EC Guide to GMP

The topic of training is defined in the EC Guide to GMP *inter alia* in Sections 2.8, 2.9, and 2.12, far exceeding the extent of the ISO 9001 regulations. In contrast to the intention of the EC Guide to GMP, however, ISO 9001 requires the systematic planning, implementation and documentation of training, even beyond the actual production and control sector.

ELEMENT XIX: SERVICING

Objective: To define regulations for customer service.

DIN ISO 9001		EC Guide to GMP
4.19	Servicing	·l.

Legend: ―― = no requirements defined, only headlines; ·l. = no similar requirement with the EC GMP Guide; the numbers given refer to the relevant requirement of the EC GMP Guide.

Content-Related Requirement

This element is generally not covered in a pharmaceutical enterprise. Element XIX deals with the sector of customer service. This element is therefore of interest for manufacturers in the medical devices sector.

ELEMENT XX: STATISTICAL TECHNIQUES

Objective: To utilize statistics as an instrument for quality control.

	DIN ISO 9001	EC Guide to GMP
4.20	Statistical techniques	——————
4.20.1	Identification of need	·l.
4.20.2	Procedures	Supplementary Guideline no. 8, Items 4 and 5. 6.12, 6.9, and indirectly 6.7, and 6.15.

Legend: —— = no requirements defined, only headlines; ·l. = no similar requirement with the EC GMP Guide; the numbers given refer to the relevant requirement of the EC GMP Guide.

Content-Related Requirement

The manufacturer is obligated to identify the need for statistical techniques for establishing and controlling the process capability and product characteristics.

Overlap with the EC Guide to GMP

The EC Guide to GMP makes reference to statistical techniques in various locations. For instance, Chapter 6.12 refers to representative samples, and 6.9 refers to trend evaluation, while Items 6.7 and 6.15 deal with the validation of test methods. Furthermore, the derivation of statistical techniques can be derived from Items 4 and 5 of Supplementary Guideline No. 8. In contrast to ISO 9000, which deals with establishing fundamentally which statistical techniques would be appropriate, the EC Guide to GMP contains detailed requirements as to where statistical techniques are appropriate and necessary.

Summary

Figure 3.3 illustrates the statements made about GMP and ISO 9000.

Figure 3.3. Quality System Documentation Model (QSDM).

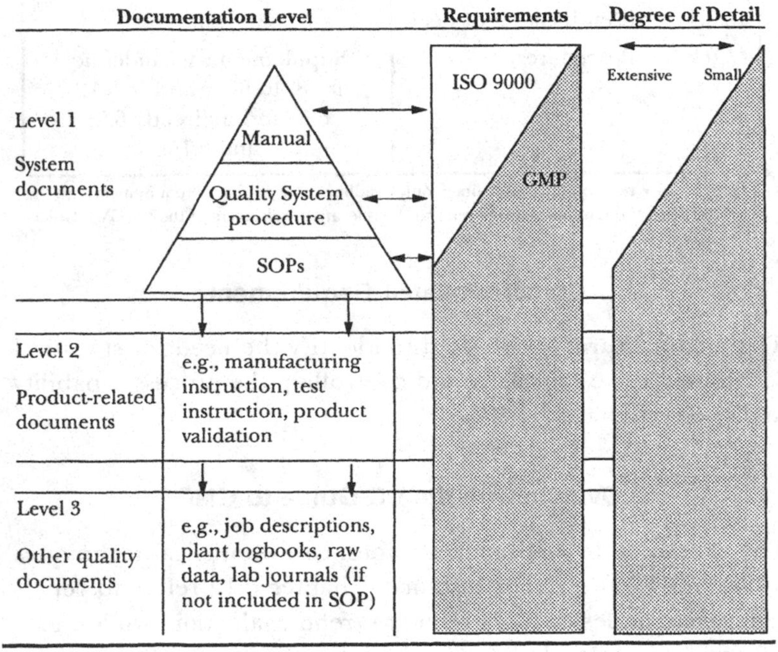

1. ISO 9000 defines requirements for the system documents, that is, for the manual, the quality system procedures and the SOPs. The degree of detail here is small.

2. GMP defines requirements for the product documents and other quality-relevant documents. Although GMP also defines system documents (e.g., SOPs), the higher one rises in the system pyramid (e.g., quality policy or manual) the less one finds in the GMP Guide.

3. The conclusion: ISO 9000 is not an alternative to GMP. It supplements it.

OUTLOOK—REVISION OF ISO 9000

The ISO 9000 series has been published as a national standard in 82 countries in the respective local languages. In over 120 countries this standards series forms the basis for the certification of quality systems. In addition, the standards series is already incorporated in EC law as EN 9000 (identical to ISO 9000), in particular by the EC Directive 90/683/EEC, the Harmonization Directive of 13.12.1990.

It is this worldwide acceptance that makes a further development of the standards series necessary. The following reasons for a revision of the standards series can be identified:

1. Since it was first published in 1987, the standards series ISO 9000 ff. has exploded in length. While in 1987 it contained 170 pages, by 1994 it had expanded to 1,050 pages (all ISO 9000 standards). For this reason it takes a specialist to come to grips with this standard.

2. There is presently no agreement in the structures of ISO 9004 (Guidelines) with the models for demonstration of conformance ISO 9001, 9002 and 9003.

3. The users of the standards series in the industrial and service sectors have called for a process-oriented standard.

4. Many companies are currently developing combined quality and environmental management systems. For this reason the users of the standard have voiced a demand for a compatability of the ISO 9000 standards series with the ISO 14000 series (environmental management).

5. Users have expressed a desire for more customer orientation and a management principle of continuous improvement.

6. In accordance with the ISO guidelines, a regular revision of the standard takes place every 5 years.

In June 1998, ISO commissioned the Technical Committees TC 20 and TC 176 to revise the standard. The planned changes are listed in the following. According to the present schedule, it is expected that the standard will be completed in 2001 and become compulsory for the supervision of quality systems in companies in 2002 at the earliest.

Some developments, however, are noticeable already:

1. The new ISO 9001 standard will no longer be structured according to the 20 "quality elements". There will be a system process model which, instead of 20 quality elements, contains 4 process steps:

   1. Leadership (management responsibility)

   2. Resource management

   3. Process management

   4. Measurement, analysis and improvement

   The process model is to be structured as shown in Figure 3.4.

2. In a purely quantitative observation it can be seen that the 4 process steps contain a total of 27 quality functions, in comparison with the previous 20 quality elements.

Figure 3.4. Quality management process model.

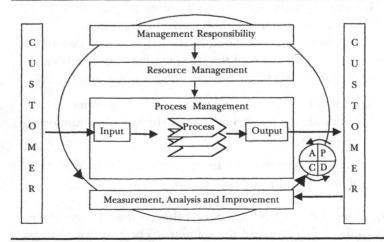

3. The quality function "Finances" of ISO 9000 will be linked to TQM models, for instance the EFQM model.

4. In addition to system audits, the new standard will include procedure audits and product audits.

5. The requirement for continuous improvement and customer orientation will occupy a broad spectrum in the new ISO 9001.

6. The structure of ISO 9001 and ISO 9004 will be identical. ISO 9001 will contain the requirements with respect to demonstration of conformity, ISO 9004 gives guidance on all the aspects of a quality management system to improve the organisation's overall quality performance. In addition, there will be compatibility with ISO 14001.

7. Many parts of the standards series ISO 9000 ff. will be deleted, for instance ISO 9002 and ISO 9003. Every company must therefore identify the elements from the new ISO 9001 which cover it and for which it wants to demonstrate conformance.

WHAT ARE THE CONSEQUENCES OF THE REVISION?

As has been mentioned, some new requirements will be added, mainly in the areas of continuous improvement and customer orientation. The ISO committee, however, will also take account of the previous 20-part structure by creating a matrix which allows the user to compare the previous 20 elements of ISO 9001 with the new structure. Even after 2002, users can employ without problems the above comparison matrix between the EC GMP Guide and ISO 9001. With the exception of the procedure and product audits, the new elements which will be contained in ISO 9001 cannot be identified in the EC GMP Guide.

On the whole, it can be said that the new process model, and thus the abandonment of functional quality elements, represents a very positive development. A process-oriented demonstration of conformance is already both advantageous and possible. At Schering Produktionsgesellschaft, this process-oriented model is already established, including the incorporation of ISO 14000.

REFERENCES

Adams, Scott. *The Dilbert Principle,* HarperBusiness. 1997.

Auterhoff, G. "Qualitätssicherung in der pharmazeutischen Industrie". *Die Pharmazeutische Industrie* (September 1991).

CDER-Human Drug cGMP Notes, December 1996.

"Der große Bluff". *Manager* (August 1995).

Frankfurter Allgemeine Zeitung (FAZ), Frankfurt/Main, September 27, 1994, Sonderteil "Qualität".

Scicchitano, Paul. 1997. *Compliance Engineering,* European Edition, "ISO goes to Washington".

4

Quality Management Systems and GMP

Lothar Hartmann
F. Hoffman—LaRoche Ltd.
Basel, Switzerland

Because the pharmaceutical industry has traditionally focused upon the application of Good Manufacturing Practice (GMP), it has been slow to consider the potential benefits to be gained by implementing an EN ISO 9001 Quality Management System (QMS).

There is still much confusion about the relationship between QMS and GMP.

Over the last few years the global pharmaceutical market has undergone significant change, forcing pharmaceutical companies to focus more than ever before on customer needs and their own internal efficiency in order to continue to compete effectively.

Today pharmaceutical companies have to satisfy a number of external requirements such as safety, environmental protection, product liability, Good Laboratory Practice (GLP), and so on. GMP is only one of these. In order to ensure reliable and efficient compliance with all these external requirements, companies need to build up a protective system for planning, organising and coordinating the necessary internal processes; that is, they have to consider establishing a QMS.

With this in mind, CEFIC's Active Pharmaceutical Systems (API) Committee commissioned the integration of current GMP requirements into the EN-ISO 9001 QMS framework. The August 1996 CEFIC/EFPIA publication "Good Manufacturing Practice for Active Ingredients Manufacturers" was combined with the relevant complementary requirements of EN-ISO 9001 "Quality Systems: Model for quality assurance in design, development, production, installation and servicing". The resulting document (CEFIC, January 1998) should be seen as a beginning for giving the pharmaceutical industry practical advice in achieving compliance with GMP requirements in a business-efficient way.

To facilitate understanding of this chapter it is important for the reader to be aware of the following points.

EN-ISO 9001 is a generic, business-focused standard which supports the effective management of quality to an internationally recognised level of best practice. It is flexible in that it specifies what is to be achieved, but allows each company freedom to determine and justify how these requirements are achieved. In contrast, GMP is an industry-specific standard prescribing what must be done to ensure product safety and efficacy. Thus, EN-ISO 9001 benefits the business by ensuring the quality of the management system, while GMP ensures that regulatory requirements are met.

Although there is inevitably some overlap between the requirements of a QMS and GMP, they are in fact highly complementary. This view is supported by a statement in the introduction to the International Inspection Convention (PIC) GMP Guideline, which refers to "a correctly implemented *system* of Quality Assurance incorporating GMP", and by the wording of the introduction in EN-ISO 9001 itself, which points out that "this international standard is complementary—*not alternative*—to the technical (product) specified requirements".

The interrelationship between EN-ISO 9001 and API GMP is illustrated in this chapter by a matrix cross-referencing the main QMS elements and GMP requirements.

GMP is not a system in its own right. It is a list of requirements presented as (mandatory) guidelines, intended to ensure product quality, safety, and efficacy. A QMS, on the other hand, is a management tool for introducing good business practice, to integrate and improve the efficiency of quality critical activities across the company, division or section.

It is incorrect to assume that systematically following GMP requirements constitutes a QMS. This view can be a source of confusion to inappropriately trained auditors.

An effective QMS has a minimum of paperwork, and should constantly question the need for the existing documents. In contrast, a bureaucratic and inefficient QMS will arise if the standard is misinterpreted and incorrectly applied.

Attempts to add on a QMS to an existing approach to GMP compliance will always fail to achieve maximum benefit. Although GMP requirements and QMS principles have much in common, they also exhibit important differences. These differences have the potential for major synergy if relevant GMP requirements can be effectively integrated into the overall system of quality management.

For the purpose of this chapter, the original EN-ISO 9001 sub-clauses have been addressed in twenty distinct chapters, supplemented by four annexes in recognition of the importance of issues concerning hygiene, facilities and utilities, validation, and change control, to the API industry. Each chapter and each of the four annexes are structured in a way that summarises the appropriate QMS principle and philosophy as a preface to the main text, which integrates relevant GMP requirements and QMS principles.

The rationale/justification and business benefits of a combined QMS/GMP approach are considered.

Safety, health and environment are not specifically addressed. However, it is widely acknowledged that implementation of a robust QMS provides a sound basis for the future development of such an Integrated Management System.

CHARACTERISTICS OF A QUALITY MANAGEMENT SYSTEM

Before identifying the characteristics of a QMS some basic definitions should be introduced:

Quality

Quality is regarded as the totality of characteristics satisfying stated and implied needs. This means that quality is not only the fulfilment of product specifications, but also takes into account the needs and expectations of customers (internal as well as external).

Quality Management

Quality management is the totality of organisational and technical measures necessary to ensure that the required level of quality is consistently achieved.

Quality Management System (QMS)

QMS is the systematic approach to, and documentation of, the organisation of Quality Management. The terms *Quality System* and *Quality Management System* may be used interchangeably.

With these explanations in mind, the fundamentals of a QMS can be described as

- Management commitment to the establishment of a quality policy.

- The quality manual as the core document outlining the QMS.

- Continuous improvement, including target setting. The main elements of this are corrective and preventive actions, management review and effective internal quality audits.

These features are essential for all QMS designed and implemented to emphasise continuous improvement, leading to increased internal efficiency.

The key prerequisites of a QMS specified in the introduction to EN-ISO 9000 standards are as follows:

- Responsibility of upper management: The management with executive responsibility shall appoint a member of its own management who, irrespective of other responsibilities, shall have defined authority for ensuring that a quality system is established, implemented and maintained.

- Resources and personnel: The company shall identify resource requirments and provide adequate resources, including the assignment of trained personnel, for management, performance of work, and verification activities including internal quality audits.

- Organisational structure: The organisational structure pertaining to the QMS should be clearly established within the overall management of a company. The lines of authority and communication should be defined.

The QMS should be defined uniquely by each company according to specific needs and ways of working. Figure 4.1 provides a typical example how this is organised in terms of documentation.

In a QMS the information flow, tasks, responsibilities and competence of staff must be clearly defined in order to achieve effective and efficient integrated structures which maximise benefits to the business.

To maintain the system, a Quality System Manager with the necessary authority and competencies should be appointed from the senior management level. Delegation of this task reflects the normal situation. In most situations this role is fulfilled by a single person, while in others cases (e.g., small companies) it may be delegated to the head of the Quality Unit, for example.

The Quality System Manager should have freedom to operate across the organisation, and authority to coordinate development, implementation and maintenance of an effective quality system.

The API manufacturer should define a quality policy to be implemented throughout the whole organisation. The quality policy should be documented, together with responsibilities and lines

Figure 4.1. Common documentation hierarchy.

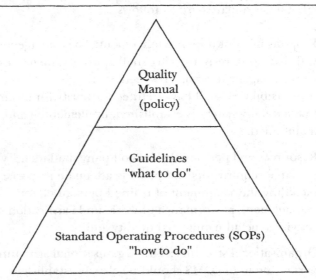

of authority, for all levels of the organisation. It should aim to prevent occurrence of nonconformities, but when they do occur, it should allow for implementation of corrective measures. The API manufacturer should provide the necessary resources and trained personnel to meet quality needs.

The QMS must be truly representative of the company's culture and way of working in order to be "owned" by the employees. If this is not achieved, the QMS will be seen as an imposed burden and will not deliver the needed benefits.

At defined intervals (at least once a year), executive management should review the adequacy and performance of the QMS to ensure that GMP and regulatory requirements, ISO quality management principles, and quality manual claims are being routinely satisfied.

The review should be based upon information gathered from the internal quality audit programme, audits by external inspectors, product reviews, trend analyses, investigations of deviations, customer complaints and any other source of relevant information. Management review should initiate actions to improve the performance of the QMS.

The QMS—in the form of organisational structure, procedures, processes and resources needed to implement the quality policy—should be described in a quality manual. The quality manual should cover relevant EN ISO 9001 and GMP requirements, describe the documentation hierarchy and indicate how the quality system is managed.

The need for quality planning is an important feature of the QMS. It requires that adequate consideration is given to an activity before implementation to reduce the risk and cost of failure.

CORRELATION OF GMP AND ISO

GMP requirements and QMS principles are not contradictory. Although there is some overlap, they are not mutually exclusive (see Table 4.1). Indeed, because of their different approaches and emphases in several important areas, they complement each other and offer significant synergies. It will, however, be a necessary challenge for the pharmaceutical industry to integrate the two in order to remain competitive in the future.

It must be stressed that the objective of EN-ISO 9000 is business benefit resulting from improved efficiency and internal and external customer satisfaction, whilst GMP is primarily concerned with consumer protection through ensuring safety and efficacy.

EN-ISO 9000 is voluntary, while GMP is mandatory.

One of the major differences between EN-ISO 9000 and GMP lies in their main emphases: ISO describes a system for managing quality, while GMP specifies how to achieve the quality of the product. GMP is industry-specific, while EN-ISO 9000 is not limited to a specific business. It provides an applicable standard to all business, from automobile manufacturers to banking (see Figure 4.2).

EN-ISO 9000 is applicable to *all* departments of a company, whereas GMP is targeted at those departments responsible for the manufacture and testing of pharmaceuticals/active ingredients (see Table 4.2).

As indicated above, there are areas where EN-ISO 9000 overlaps with GMP (e.g., production and documentation). In these cases the more prescriptive requirements of GMP complement the generic character of the EN ISO 9000.

Table 4.1. Comparison of GMP and ISO requirements.

Comparison ISO/GMP		
○	1.	Management responsibility
○	2.	Quality system
○	3.	Contract review
○	4.	Design control
●	5.	Document data control
○	6.	Purchasing
○	7.	Control of customer-supplied product
●	8.	Product identification and traceability
●	9.	Process control
●	10.	Inspection and test status
○	11.	Control of inspection, measuring, and test equipment
●	12.	Inspection and test status
●	13.	Control of nonconforming product
○	14.	Corrective and preventive action
●	15.	Handling, storage, packaging, preservation, and delivery
●	16.	Control of quality records
○	17.	Internal quality audits
○	18.	Training
○	19.	Servicing
○	20.	Statistical techniques

Key
● GMP is more detailed than ISO.
○ ISO requires more than GMP.

A matrix at the end of this chapter illustrates how the product requirements (GMP) correlate with the system requirements (ISO). There are no items where the GMP and ISO requirements are contradictory.

RATIONALE

Management Responsibility (Chapter 1)

To be effective, the QMS must have the visible and ongoing support of top management.

Figure 4.2. Different approaches to GMP requirements and QMS principles.

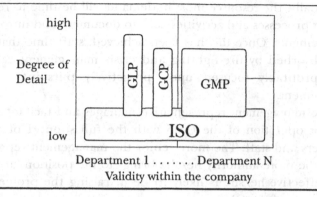

high

Degree of Detail

low

Department 1 . . . . Department N
Validity within the company

To fully benefit the company, the QMS should involve all staff whose activities influence quality, have a clear and unambiguous focus on continuous improvement and incorporate relevant, realistic performance measures with emphasis on reducing failure costs and satisfying internal and external customer needs.

The quality manual occupies the highest level in the document hierarchy. It overviews and acts as a directory to the QMS, capturing the unique character of the company.

Maximum benefit from a QMS will only be achieved if top management support is clearly and repeatedly visible.

A clearly defined organisational structure with unambiguous responsibilities and authorities is essential to avoid confusion,

Table 4.2. Comparison of EN-ISO 9000 and GMP.

	ISO	GMP
Objectives	Customer satisfaction	Customer protection
Binding Nature	Voluntary	Legally binding
Monitoring	Neutral institution	Authority; prescribed by law
Degree of detail	General, all departments	High; especially production, QA/QC
Type of business	Neutral all types	Specific for pharmaceutical
Main emphasis	System-related	Product-related

duplication and/or omission and the consequent increased risk of quality failure.

Initially, the resource most in demand will be time to review existing processes and activities, and to document and introduce improvements. Once this has been achieved, staff time that was once absorbed by fire-fighting and crisis management can be more profitably focused upon proactively pursuing system improvement.

The management representative manages and facilitates the efficient operation of the QMS with the full support of other managers and staff. The more senior the management representative, the more authority he/she brings to the position, and the more effective he/she is likely to be in raising the profile and importance of the QMS.

All successful organisations are dynamic in that they change constantly in response to internal and external pressures. This is especially so if a policy of continuous improvement is encouraged.

The management review requirement is the opportunity for senior management to review performance of the QMS against quality manual claims and organisation needs. Management review, together with other activities such as internal audits and corrective and preventative action, is a key driver of continuous improvement.

Quality System (Chapter 2)

If the quality system is not adequately documented, then for all practical purposes it does not exist, since there can be no assurance of process (and therefore product) consistency. Care must be taken, however, to ensure that the quality system is not excessively bureaucratic. It is important to establish a realistic balance between documented procedure, training and experience. The level of detail in written procedures should be such that any less detail would adversely affect product quality.

The quality manual instructs employees and informs customers, external auditors and other relevant parties that the organisation has an efficient and cost-effective means of managing quality. It provides a summary of, and directory to, the quality system. It will, of necessity, encompass the ISO 9000 framework, but in doing so should realistically reflect the unique culture, processes, and ways of working of the organisation. Off-the-shelf quality manuals should be avoided at all costs.

Systematic forward planning is an important factor in the avoidance of expensive quality failures. The quality manual may be considered as the master quality plan.

Contract Review (Chapter 3)

This refers to agreements between the API manufacturer and the customer. It requires that the API manufacturer be confident, before the order is accepted, that customer needs can be satisfied.

A clear understanding of what the customer wants is essential to ensure that expectations are met. It is important, therefore, that a robust communication interface exists not only between the customer and the sales department, but also between sales and other relevant parts of the organisation. Failure of these communication interfaces could lead to a misunderstanding of the customer's requirements, resulting in failure to satisfy customer needs and potentially lost business.

Design Control (Chapter 4)

In today's increasingly competitive market, a delay in launching a new product and/or a major development overspend is likely to significantly reduce lifetime profits for that product. This is one reason why the design and development process should be carefully planned and controlled. Another is that the quality of the product should be right the first time. The control mechanism has to be robust enough to cope with the need for change inherent in any development process. Adequate control is exercised through development milestone meetings to review and assess progress for approval before proceeding to the next phase of development, verify that product and process characteristics satisfy approved specifications, and ensure that customer requirements are going to be met.

A well-controlled design and development process will facilitate the on-time, within-budget introduction of high-quality new products.

It should also be remembered that the principles of design control may apply when changes to an existing product are required. The extent to which design control is applied will be determined by the nature and complexity of the change.

Document and Data Control (Chapter 5)

The quality system must be documented to ensure that all users have a common and uniform understanding of its function. An effective mechanism for controlling this documentation is necessary, since it is essential that everyone concerned has access to the most recent version. Failure to maintain up-to-date documentation, or to have it available when and where it is needed, is a common cause of nonconformity and associated failure costs due to rejection, rework, failure investigation and the potential for loss of confidence by customers and/or regulatory authorities.

Document control is assisted by keeping documentation as brief and concise as possible.

Purchasing (Chapter 6)

It is essential that quality critical materials and services (such as equipment calibration, pest control, off-site storage and distribution) are only purchased from reliable sources. Failure to clearly and fully communicate purchasing requirements and failure to ensure that these sources have the capability to satisfy quality expectations, are common causes of potentially expensive downstream quality problems. In contrast, a close and carefully controlled working relationship with a supplier of proven quality can significantly reduce overall quality-related costs.

Customer-Supplied Product (Chapter 7)

The API manufacturer may, from time to time, be required to handle (receive, process, store, despatch), and therefore be responsible for, material which it does not own. Examples may include the use of special packaging supplied by the customer, or the "contracting in" of activities (grinding, milling, micronising of bulk) at the request of another company. It is important that these issues are considered during contract review, and that elements of the API manufacturer's quality system are capable of ensuring that the quality of this customer-supplied product is not impaired.

Identification and Traceability (Chapter 8)

The ability to trace the history of materials and activities influencing quality is an essential feature of the API manufacturer's Quality System. Not only is it necessary for effective overall management, it is particularly important in identifying the root cause of quality problems and therefore fundamental to achieving an effective and lasting solution. An efficient identification and traceability procedure ensures rapid location of suspect material and facilitates action to prevent the incursion of further failure costs.

Process Control (Chapter 9)

Effective process control is central to the concept of a system for managing quality. Management should therefore be able to demonstrate that control has been established through the existence of documented planning, training, supervision and monitoring procedures.

A clear understanding of each process, and its interface with other relevant processes, is a prerequisite for overall control.

The GMP/ISO 9001 CEFIC/APIC Guideline, referred to in the introduction, has extended the chapters given by EN-ISO 9000 to separately address the topics of hygiene (chapter 21); facilities, utilities and engineering (chapter 22); validation (chapter 23); and change control (chapter 24), in view of their importance to the API manufacturers.

Inspection and Testing (Chapter 10)

Even with today's highly sophisticated technology, no process is 100 percent reliable, particularly when part of that process involves some degree of human intervention. It is therefore necessary to plan and perform inspection and test activities at appropriate stages of a process if the customer is to consistently receive API of specified quality.

Control of Inspection, Measuring and Test Equipment (Chapter 11)

To ensure that processes are under control, they need to be monitored (inspected and/or tested), which involves some form of measurement. Measurements are only meaningful if they can be relied upon, and they can only be relied upon if they are obtained using instruments of known accuracy. Without this level of assurance, processes cannot be considered to be under control.

Inspection and Test Status (Chapter 12)

The effort and expense associated with inspection and testing are largely wasted if the outcome is not readily accessible, or specifically communicated, to those who need to know. An effective means of quickly and clearly identifying the quality status of incoming, in-process and finished goods is essential for efficient and cost-effective operation. It helps to ensure that only satisfactory material is processed, and only quality goods are supplied to the customer. In contrast, failure to adequately identify quality status can prove to be very expensive if it leads to reprocessing, rejection or recall.

Control of Nonconforming Product (Chapter 13)

Defective product should not be released for supply to the customer. Most defects are the result of a weakness in part of the quality system. Therefore, the occurrence of nonconforming product should be viewed as an opportunity to improve the existing quality system.

Failure to control nonconforming product, to fully investigate its cause, or to identify and implement appropriate improvement to the quality system, will inevitably have an adverse impact upon the organisation in the form of unnecessarily high failure costs due to rework/reject/recall and potential loss of business.

Corrective and Preventive Action (Chapter 14)

Corrective and preventive action is intended to eliminate the cause(s) of quality system failures as, or before, they arise.

Together with management review and internal quality audit, corrective and preventive action is one of the main drivers of continuous improvement.

A planned, structured and coordinated approach to this aspect of the QMS greatly increases the likelihood of identifying the true cause of a quality problem, and of identifying and implementing an acceptable and lasting remedy. A less structured approach is more likely to result in an unsatisfactory, possibly temporary, solution to a quality problem.

Proactive remedial action which highlights areas of concern (potential weaknesses in the quality system) and initiates action before that weakness results in a quality failure, is likely to be particularly beneficial. Prevention is more cost-effective than cure.

If the corrective and preventive action part of the quality system is not working effectively, failure costs will be higher than necessary and, more importantly, opportunities to improve the quality system may be lost.

The benefits from remedial action should be captured and quantified by measuring improved performance wherever it is practical to do so.

Handling, Storage, Packaging, Reservation and Delivery (Chapter 15)

This chapter emphasises the importance of protecting product quality during manufacture and prior to receipt by the customer. Inappropriate handling, incorrect storage, transit damage and so on, can easily negate benefits derived from the rest of the quality system.

Control of Quality Records (Chapter 16)

Whether retained as hard copy or electronic documents, or as physical samples, quality records are a vital source of objective evidence for the effectiveness (or otherwise) of the organisation's quality system. An efficient records management procedure will be most appreciated during audit and investigation situations, when timing is often a high priority. In contrast, an ineffective procedure which loses records can be an expensive liability.

Virtually every activity generates a record of some type. The key to controlling quality records efficiently is to begin by identifying which types of records should be retained and for how long, that is, which records are useful and which are not.

Internal Quality Audits (Chapter 17)

A comprehensive internal quality audit programme undertaken by well-trained, knowledgeable auditors will provide an essential healthcheck of the quality system and its day-to-day operation. It will also provide a high-profile demonstration of management's commitment to the quality policy and the claims of the quality manual.

Internal quality audits are a primary means of identifying areas of the quality system in need of attention. Together with corrective and preventive action (the means for improvement), and the authority afforded by management review, internal quality audits are at least a means of preventing deterioration of the quality system, and ideally a driving force for continuous improvement.

The benefits of an internal quality audit programme are completely wasted if resulting actions are not satisfactorily completed.

Training (Chapter 18)

Adequate training is a regulatory necessity. In addition to this basic requirement of the API industry, it is logical to assume that well-trained staff will be better motivated, more efficient, and less likely to make expensive mistakes, thereby contributing more to the organisation's overall success.

As competency requirements change and evolve, a robust training policy is invaluable for improving staff skills and maintaining a competitive edge.

Servicing (Chapter 19)

Although after-sales service/support is an important stand-alone feature for many industries, this is not usually the situation with API manufacturer, since issues such as customer complaint handling are more appropriately dealt with in other chapters.

Statistical Techniques (Chapter 20)

Because all processes and measurements are subject to variation, the sensible application of statistical analysis increases the probability of correct decisions by translating data into meaningful information.

The correct application of appropriate statistical techniques leads to a better understanding of process capability and facilitates more cost-effective operation.

Hygiene (Chapter 21)

The implications for patient health of using contaminated medicines are self-evident. Commercial implications for the API manufacturer could be critical if inspection findings and/or reasons for product withdrawal/recall cause loss of confidence by regulatory authorities and customers.

Facilities, Utilities, Engineering (Chapter 22)

Unsuitable manufacturing conditions and inadequate engineering support will significantly increase costs through reduced efficiency, increased risk of batch rejection and failure to satisfy regulatory authorities and customer expectations.

Validation (Chapter 23)

Understanding and characterisation of process capability is an important factor in identifying areas for improving robustness and establishing confidence in manufacturing processes. Validation of both manufacturing and analytical methods commands an increasingly higher profile during regulatory inspections, particularly where computerised systems are employed.

Change Control (Chapter 24)

Change is inevitable if there is to be improvement. It must, however, be planned, controlled and coordinated if chaos and confusion are to be avoided. Failure to adequately control change will greatly increase the risk of incurring failure costs due to situations such as missed new product introduction deadlines; use of outdated documentation; and failure to keep up to date with regulatory and customer requirements. Staff morale and motivation will suffer, as well as will management credibility.

PERFORMANCE MEASUREMENTS

The API industry is constantly changing. It is driven by the influence of technological advances and constrained by regulatory requirements, global market forces and local market pricing considerations. Setting and achieving medium- and long-term objectives is therefore not easy. When failures occur, the true underlying cause(s) must be established if learning points are to be identified and appropriate corrective measures applied.

The EN-ISO 9001 QMS approach requires the adoption of a corporate philosophy of continuous improvement leading to greater efficiency and increased customer satisfaction. Although the implementation of an effective QMS is complex, if done correctly and successfully it provides a sound basis for progress toward excellence. Experience has shown that once a QMS has been introduced, there is a risk that the pace of ongoing improvement generated by the introduction of the QMS may falter and lose momentum. It is for this reason that an objective evaluation of improving performance is essential. A number of possibilities exist.

At the strategic level, the corporate goal may be to achieve recognition through formal ISO certification as a stepping stone to gaining the European Quality Award or the U.S.-based Malcolm Baldridge Award. Each involves detailed evaluation of systems performance which in itself allows management to redefine processes that fail to meet performance objectives.

At the tactical level, an integral part of successfully implementing an effective QMS is the need to identify, agree upon, and use realistic criteria for routinely monitoring performance trends. Some general examples are provided below. The nature and emphasis of performance measures will inevitably vary from one company to another.

Management Responsibility

- Frequency of management review meetings actually held (versus scheduled)
- Number of improvement initiatives generated and implemented

Quality System

- Incidence of nonconformity
- Relative categories of nonconformances
- System improvement initiatives in progress or completed

Contract Review

- Customer satisfaction score
- Delivery against agreed lead time

Design Control

- Project delays
- Significant post-launch quality specification revisions
- Quality-related cost overruns

Document Control

- Proportion of controlled documents beyond official review date
- Delays/failure costs due to lost or out-of-date documentation
- Relevant internal quality audit observation trends

Purchasing

- Late delivery
- Percentage of deliveries with quality problems
- Customer complaints traceable to purchased material/services

Change Control

- Percentage of controlled documents beyond review date
- Overall effectiveness of change-control procedure (score based on users' views)
- Delays due to weaknesses in change-control procedure

Control of Customer-Supplied Product

- Instances of failure to control quality of customer-supplied product

Identification and Traceability

- Impact upon complaint response time
- Lost or misplaced material/product

Process Control

- Rejected, reworked, reprocessed, or recalled product (incidence/cost)

Inspection and Testing

- Failure to detect defects, leading to subsequent quality problems

Control of Inspection, Measuring and Test Equipment

- Instances of measuring instrumentation overdue for calibration
- Delays and retests due to equipment being out of calibration

Inspection and Test Status

- Examples of inadequate status identification
- Quality failures which could have been avoided by adequate status identification

Control of Nonconforming Product

- Process deviations per batch
- Reworks per month
- Cost of rejects per month

Corrective and Preventive Action

- Preventive actions ongoing/completed
- Corrective actions ongoing/completed
- Number of customer complaints
- Customer complaint turnaround time

Handling, Storage, Packaging, Preservation and Delivery

- Damage to product during handling or storage in-house
- Instances of transit damage

Quality Records

- Average time to locate archived records
- Number of records lost or misplaced
- Records retained beyond official destruction date

Internal Quality Audits

- Percentage of audits performed on schedule
- Average major/critical observations per area audited
- Percentage of actions completed within agreed timeframe
- Observations by external auditors

Training

- Percentage of staff with current training need identified/outstanding
- Percentage of training records available/up to date

Statistical Techniques

- Use of data analysis
- Statistically based sampling plans
- Instances of statistically based experimental investigation design
- Training in application of statistical techniques

Hygiene

- Levels of product contamination
- Customer complaints of foreign matter or microbial contamination
- Cleaning process failures

Facilities, Utilities, Engineering

- Equipment breakdown
- Quality trends involving air, water, and so on
- Security breaches
- Cross-contamination
- Reprocessing due to inadequate equipment performance and/or environmental conditions

Validation

- Proportion of quality critical processes validated or in need of validation
- Proportion of quality critical analytical methods validated or in need of validation

BENEFITS

EN-ISO 9000 has achieved recognition as an internationally acknowledged quality standard. If applied correctly and sensibly, the common sense contained within its pages has the potential to generate significant and sustainable business benefits through improved performance, reduced failure costs and increased competitiveness. If implemented correctly, it ensures that employees obtain a better understanding of quality principles and become comfortable working in a cross-functional environment where old departmental barriers are overcome.

From a business perspective, the potential benefits to all stakeholders of an effective QMS may be grouped as follows.

Customer Focus

A clear understanding of each customer's current needs is essential to ensure that quality requirements are met. It is important, therefore, to have in place a robust communications network with effective interfaces between not only Marketing/Sales and the customer, but also between Marketing/Sales and other relevant departments within the organisation in order to achieve full and effective contract review.

The changing needs of customers present a constant challenge to any company whose objective is to remain the supplier of choice. In today's increasingly competitive marketplace, delay in launching a new product and/or a major developmental overspend can significantly reduce the lifetime profits of a new product. To meet this challenge the design and development process must be carefully planned and controlled by a system which must be flexible enough to cope with the changes inherent in any development process, yet robust enough to ensure that stringent time and budget targets are achieved.

Continuous Improvement Focus

If the QMS is designed and implemented to emphasise continuous improvement (driven by the effective use of internal quality audits, corrective and preventive action and management review), then internal efficiency will rise, leading to a sustainable reduction in failure costs. Similarly, the effective control of nonconforming product helps to identify the basic cause of quality problems, and in so doing provides an important improvement opportunity.

A comprehensive internal quality audit system is a vital healthcheck and provides a means of identifying issues in need of attention, while a planned and structured approach to corrective and preventive action increases the likelihood that the basic cause of actual or potential quality problems will be identified and lasting remedial action taken.

A well-trained workforce is likely to be better motivated, more efficient, and less likely to make expensive mistakes. As competency requirements change and evolve, an efficient training system is an invaluable contribution to any company's competitive edge and overall success.

Experience in a variety of industries has confirmed the value of an EN-ISO 9001–based QMS as a firm foundation for progressing to world-class status, as exemplified by achievement of the European Quality Award and equivalent. It also provides a suitable springboard for developing an Integrated Management System incorporating quality, safety and environment.

Regulatory Compliance and (Continuous) Inspection Readiness

It is common for a company to commit a vast number of costly personnel-hours, and expose staff to the associated stress and disruption, in preparing for inspection by a regulatory authority. Because such inspections look for evidence of GMP compliance, the ability to demonstrate effective control through a documented QMS will help demonstrate to the inspecting authority that all relevant aspects of product quality have been addressed.

Since many GMP deficiencies are the result of a weakness in, or failure of, part of the QMS, an effective internal quality audit system will go a long way toward ensuring regulatory compliance, and will facilitate continuous inspection readiness.

Recognised Best Practice

Compliance with this internationally recognised business standard for managing quality, whether through third-party certification or a less formal in-house approach, will provide visible evidence to regulatory authorities, customers, and staff alike that a company's processes are under control. The strong emphasis on forward planning will minimise expensive quality failures. A realistic practical balance between the level of documented procedures, training, qualifications and experience will ensure a minimum of bureaucracy. The level of documentation should not exceed that required to maintain the desired level of quality.

Reduced Stress

A concisely documented QMS, having the full visible support of top management, will lead to better understanding of employee roles, responsibilities, authorities, and working interfaces. It will avoid confusion, and reduce the risk of omission and/or duplication. Less staff time will be absorbed by fire-fighting and crisis management, allowing more time to be devoted to improving operating efficiency.

CERTIFICATION

Formal certification provides an independent measure of conformity to QMS principles.

Certification itself is not mandatory. A company may operate a QMS without certification.

The pressure for formal certification has been a factor damaging the image of the QMS in the past. Other factors include the mistaken belief that absolutely everything must be documented, and the widely varying competencies of QMS consultants and, to a lesser extent, of companies accredited to assess a QMS and provide certification. Careful selection is therefore essential if maximum benefit (and value for money) is to be obtained (Figure 4.3).

The achievement of formal certification is a means of officially acknowledging that the QMS is of internationally acceptable standard. It also helps to ensure that this standard is maintained and (hopefully) developed through continuous improvement to reach a higher level of efficiency.

Figure 4.3. Certification acknowledges a standard.

CONCLUSION

EN-ISO 9000, the world's best-known quality system, is well established and subject to numerous comments, which makes it an international standard that is undergoing continuous improvement. For this reason, it is appropriate for the pharmaceutical industry to adopt the EN-ISO 9000 framework incorporating the current GMP requirements.

EN-ISO 9000 will not—and cannot—replace GMP, and vice versa. This was never the intention.

A poor understanding of the QMS concept will lead to an excessively bureaucratic system which employees will not accept.

It is amazing that some pharmaceutical companies still rely on the outdated view that they cannot and do not need to implement a QMS because they have to comply with GMP regulations. This short-sighted lack of understanding only serves to delay the benefits to be derived from integrating GMP into the QMS framework.

MATRIX

The matrix in Table 4.3 cross-references the major GMP requirements as laid down in CEFIC/APIC GMP guideline with the relevant chapters decribed in EN-ISO 9000.

Table 4.3. Major GMP requirements.

ISO Chapters	1	2	3	4	5	6	7	8	9	10	11	12	13	14	15	16	17	18	19	20	21	22	23	24
Organisation, Personnel, and Training																								
Organisation																								
Management commitment				X																				
Organisation				X														X						
General Personnel and Training																								
Sufficient personnel				X																				
Training																		X						
Hygiene																								
Infectious disease																					X			
No contact with API																					X			
Clothing																					X			
Facilities and Utilities																								
Buildings																								
Design																						X		
Outside equipment																						X		

ISO Chapters	1	2	3	4	5	6	7	8	9	10	11	12	13	14	15	16	17	18	19	20	21	22	23	24
Construction																						X		
Handling of API																						X		
Separate areas and equipment															X							X		
High actives																						X		
Other facilities																						X		
Laboratory areas																						X		
Ventilation																								
Design																						X		
Filters																						X		
Recirculated air																						X		
Air intake points																						X		
Utilities and Services																								
Specifications																						X		
Critical utilities																						X		
Water quality																						X		
Potable water																						X		
Treated water																						X		
Endotoxin-free water																						X		

Continued on the next page.

Continued from the previous page.

ISO Chapters	1	2	3	4	5	6	7	8	9	10	11	12	13	14	15	16	17	18	19	20	21	22	23	24
Heating or cooling water																						X		
Steam																						X		
Other utilities																						X		
Pipework																								
Risk of contamination																						X		
Labelling																						X		
Waste Disposal																								
Identification of containers																						X		
Compliance with local laws																						X		
Security																								
Prevention of unauthorised access																						X		
Storage Facilities																								
Availability and conditions															X									
Storage in the open															X									
Separate storage													X											

ISO Chapters	1	2	3	4	5	6	7	8	9	10	11	12	13	14	15	16	17	18	19	20	21	22	23	24
Cleanliness and Hygiene																								
Facilities															X						X			
Restrictions																					X			
Written procedures																					X			
Equipment and Production																								
Equipment																								
Design																						X		
Non-reactivity																						X		
Closed equipment																						X		
Suitability for intended use																							X	
Labelling																						X		
Cleanliness of Equipment																								
Product contact surfaces																						X		
Cleaning procedures																						X		
Cleaned equipment																						X		
Level of cleanliness										X												X	X	
Cleaned at intervals																						X		

Continued on the next page.

Continued from the previous page.

ISO Chapters	1	2	3	4	5	6	7	8	9	10	11	12	13	14	15	16	17	18	19	20	21	22	23	24
Control of Product Contamination																								
Final Product Stages									X															
Measures to avoid contamination									X															
List of measures									X															
Isolation, Drying, Blending									X															
Special care									X															
High actives									X															
Selective blending									X															
Control, Monitoring, and Testing Equipment																								
Maintenance and calibration											X													
Investigation of deviations											X													
Computerized Systems																								
Validation																								
Extent of validation																						X		
Prospective validation																						X		X
Retrospective validation																							X	

ISO Chapters	1	2	3	4	5	6	7	8	9	10	11	12	13	14	15	16	17	18	19	20	21	22	23	24
Software																								
Categories to be validated																						X		
Changes																								
Alterations and modifications																						X		X
Temporary changes																								X
Testing																								
Activities																						X		
Security																								
Identifying persons																						X		
Critical data																						X		
Backup system																						X		
Documentation																								
SOPs																						X	X	
Ancillary Aspects																								
Incident reporting																						X		

Continued on the next page.

Continued from the previous page.

ISO Chapters	1	2	3	4	5	6	7	8	9	10	11	12	13	14	15	16	17	18	19	20	21	22	23	24
Documentation																								
General Considerations																								
Carefully prepared					X																			
Drawn by knowledgeable persons					X																			
Compliance					X																			
Entries					X																			
Manual entries					X																			
Specifications and Test Procedures																								
Availability										X														
Written test procedures										X														
Critical tests validated										X													X	
Test Records																								
Records of tests										X														
Independent check										X														
Retention of records																X								
Production Documentation																								
Production overview									X															

ISO Chapters	1	2	3	4	5	6	7	8	9	10	11	12	13	14	15	16	17	18	19	20	21	22	23	24
Instructions										X														
Instructions for packaging										X														
Instructions for other activities										X														
Batch record										X														
Electronic batch record																X								
Evaluation of deviations										X														
Packaging to be recorded										X														
Batch record review										X														
Retention of documents										X														
Validation																								
Validation Policy																								
Written procedure																						X		
Preliminary Considerations																								
Start of validation				X																				
Study				X																				
Qualification																								
Qualification of equipment																						X		

Continued on the next page.

Continued from the previous page.

ISO Chapters	1	2	3	4	5	6	7	8	9	10	11	12	13	14	15	16	17	18	19	20	21	22	23	24
Activities																							X	
Process Validation																								
Prospective validation																							X	
New API																							X	
Number of batches																							X	
Retrospective validation																							X	
Concurrent validation																							X	
Scope																								
Critical steps																							X	
Critical process parameters																							X	
Critical quality attributes																							X	
Variables to be considered																							X	
Validation Documentation																								
Validation Plan																							X	
Validation Work																							X	
Deviation review																							X	
Malfunctioning cause																							X	

ISO Chapters	1	2	3	4	5	6	7	8	9	10	11	12	13	14	15	16	17	18	19	20	21	22	23	24
Cause not identified																							X	
Validation report																							X	
Archiving																							X	
Revalidation																								
Changes to processes																							X	
Periodic evaluation																							X	
Change Control																								
Change Control Procedures																								
Written procedure																								X
Aspects to be considered																								X
Approval of changes																								X
Implementation of Changes																								
Implementation																								X
Evaluation after change																								X
Contract Manufacture or Analysis																								
Contract																								
Written and approved contract			X																					

Continued on the next page.

Continued from the previous page.

ISO Chapters	1	2	3	4	5	6	7	8	9	10	11	12	13	14	15	16	17	18	19	20	21	22	23	24
Recommendations on GMP			X																					
Materials Management																								
Purchasing and Control																								
Purchase against specifications					X																			
Reduction of in-house testing					X																			
Evaluation of suppliers					X																			
Changing suppliers																								X
Receipt and Quarantine of Materials																								
Written procedures										X														
Solvents										X														
Non-dedicated tank trucks										X														
Delivery of more than one lot										X														
Quarantining until decision															X									
Measures before mixing										X					X									
Retesting period for raw materials										X														
Storage of labels										X														
Separation of rejected materials												X	X	X										

ISO Chapters	1	2	3	4	5	6	7	8	9	10	11	12	13	14	15	16	17	18	19	20	21	22	23	24
Issuing or Distribution of Materials																								
Use after testing										X														
Measures if used before testing										X														
Subdivision								X																
Specific Management of APIs																								
Storage of APIs														X										
Measures for distribution										X														
Records of distribution								X																
Sampling																								
Facilities for Sampling																								
Criteria for facilities																						X		
Additional criteria																						X		
Sampling Procedures																								
Sampling tools									X															
Written procedures									X															
Hazardous materials						X																		
Sample containers														X										

Continued on the next page.

Continued from the previous page.

ISO Chapters	1	2	3	4	5	6	7	8	9	10	11	12	13	14	15	16	17	18	19	20	21	22	23	24
Number of samples										X														
Sampling plan										X														
Measures to sample APIs															X									
Filling and Labelling of APIs																								
Packaging Materials																								
No effect on product quality															X									
Filling/Packaging																								
Environment																						X		
Avoiding cross-contamination									X															
Directions for filling															X									
Reusable containers															X									
Labels and Labelling																								
Control of labels								X																
Labelling of containers								X																
Special requirements								X																
Containers used in production								X																
Reusable containers								X																

ISO Chapters	1	2	3	4	5	6	7	8	9	10	11	12	13	14	15	16	17	18	19	20	21	22	23	24
Engineering																								
Compliance with GMP																						X		
Affected areas																						X		
Change control																								X
Quality Management																								
Quality Assurance																								
Philosophy		X																						
Quality assurance system		X																						
Evaluation of API										X														
Investigation of complaints														X										
Regular quality reviews		X																						
Auditing																	X							
Quality Control																								
Responsibilities										X														
Expertise										X														
Disposal										X														
Documentation						X																		

Continued on the next page.

Continued from the previous page.

ISO Chapters	1	2	3	4	5	6	7	8	9	10	11	12	13	14	15	16	17	18	19	20	21	22	23	24
Contract analysis			X																					
In-Process Control (IPC)																								
Approval										X														
Points described in writing										X														
Limits described in writing										X														
Test procedures										X														
Equipment used											X													
Results										X														
Rejection, Recovery, Reprocessing, and Returns																								
Rejection of Materials																								
Adequate storage													X											
Additional controls before use													X											
Use of nonconforming materials													X											
Evaluation of affected batches													X	X										
Recovery of Materials and Solvents																								
Recovered materials									X															

ISO Chapters	1	2	3	4	5	6	7	8	9	10	11	12	13	14	15	16	17	18	19	20	21	22	23	24
Recovered solvents								X																
Mother liquors									X															
Reprocessing or Reworking of Materials																								
Occasional reprocessing													X											
Rework													X											
Procedures													X											
Documentation													X											
New batch number													X											
Returned Materials																								
Quarantining										X														
Decision on how to handle										X														
Stability Testing and Retest Date																								
Storage Conditions and Retest Date																								
Conditions				X											X									
Retest date				X																				
Different retest dates				X																				

Continued on the next page.

Continued from the previous page.

ISO Chapters	1	2	3	4	5	6	7	8	9	10	11	12	13	14	15	16	17	18	19	20	21	22	23	24
Specification				X																				
Stability Testing																								
Validated test procedures				X																				
Storage conditions for samples				X																				
Monitoring				X																				
Production at different sites				X																				
Potential effects of changes				X																				
Evaluation of data																				X				
Complaint and Recall Procedures																								
Complaint Procedures																								
Registration														X										
Review														X										
Records														X										
Recall Procedures																								
Written procedure													X	X										
Evaluation of information													X	X										
Notification of authorities													X	X										

ISO Chapters	1	2	3	4	5	6	7	8	9	10	11	12	13	14	15	16	17	18	19	20	21	22	23	24
Self-Inspections																								
Regular																X								
Recording of findings																X								
Responsibility for correction																	X							
Retention Periods																								
Documentation																								
Written procedures									X							X								
List of documents									X							X								
Other documents									X							X								
Samples																								
Raw materials									X							X								
APIs									X							X								
Purpose of retaining									X							X								
Use for other purposes																								

ACKNOWLEDGMENTS

The author of this chapter wants to thank these people for their valuable contributions and discussions: Ms. Lucienne de Morsier, Novartis, Switzerland; Dr. Robert Hopkins, GlaxoWellcome, U.K.; and Dr. Arnd Hardtke, Arthur D. Little Inc., Germany.

REFERENCES

CEFIC. January 1998. "Quality Management System for API Manufacturers integrating GMP into ISO 9001".

CEFIC/EFPIA. August 1996. Guideline "Good Manufacturing Practice for Active Ingredient Manufacturers".

EN ISO 9001. July 1994. "Quality Systems—Model for quality assurance in design, development, production, installation and servicing".

Hopkins, Robert. 1998. "Quality Management System for APIs". *PharmTech Europe* 10 (6):49.

Schmidt, Oliver. 1997. "Process orientation in Quality Assurance" Part I and II, *PharmTech Europe* 9 (10):40–44 and 9 (11):36–44.

5

The Master Plan Concept: GMP-Compliant Production in a Non-GMP Environment

Karl Metzger
Concept Heidelberg,
Heidelberg, Germany

Starting materials for the production of pharmaceuticals are often obtained from large chemical plants which usually produce "technical quality". The amount used in pharmaceutical production is often far below 1 percent of the total production. In some cases these large chemical plants are certified according to the International Organization for Standardization (ISO) 9000 series, while in others there is no quality system installed. The question is how to produce the "pharmaceutical grade" products in a Good Manufacturing Practice (GMP)-compliant environment. In this chapter, two methods of incorporating GMP in non-GMP environments are shown. Companies which already have a quality system can supplement their quality manual, either by adding one chapter at the end of the manual, or by expanding the chapter "Process Control". Another approach is the "Master Plan Concept", which can be applied to firms with or without an existing quality system. In this approach, all GMP requirements are laid down in "Master Plans".

CURRENT SITUATION

The European Commission (EC) Directive, which sets forth the principles and guidelines of GMP for medicinal products for human use (91/356/EEC), states in Article 6, "Quality Management", "The manufacturer shall establish and implement an effective pharmaceutical quality assurance system, involving the active participation of the management and personnel of the different services involved". In the Code of Federal Regulations (CFR), the U.S. counterpart of EC Directives, there is no explicit requirement for such a quality system, but companies must nonetheless establish a system which assures the production of quality.

Industry's classic approach to implementing a quality system is that described by the ISO 9000 series. Most companies which built up a quality system used these standards as a guideline for the installation of the system. Furthermore, with a system arranged in the order of the 20 chapters of ISO 9001 (1994a), as shown in Table 5.1, it is easy to show auditors that all requirements for certification according to the applied standard are fulfilled.

Table 5.2 shows that the main emphasis of GMP varies from the topics required by ISO 9001.

It makes no sense to build up two different quality systems in one company to fulfill the requirements of ISO and GMP; these requirements should be integrated in one system. In the literature, quality systems are mentioned which integrate both ISO and GMP in one quality system (Schmidt 1997a, b). Table 5.3 shows an example of a quality system arranged in a process-oriented order. This system can be extended to include financial considerations of quality, according to ISO 9004 (1994b), and environmental management systems with regard to ISO 14001 (1996) or EMAS (1996).

Drafts from the EC (1998a) and the Food and Drug Administration (FDA) (1998b) will lead to GMP guidelines which set

Table 5.1. Classical structure of a QMS according to ISO 9001 (1994).

Chapter	Topic
0	Introduction
1	Management responsibility
2	Quality system
3	Contract review
4	Design control
5	Document and data control
6	Purchasing
7	Control of customer-supplied products
8	Product identification and traceability
9	Process control
10	Inspection and testing
11	Control of inspection, measuring and test equipment
12	Inspection and test status
13	Control of nonconforming products
14	Corrective and preventive actions
15	Handling, storage, packaging, preservation and delivery
16	Control of quality records
17	Internal quality audits
18	Training
19	Servicing
20	Statistical techniques

forth certain requirements for pharmaceutical starting materials. This will cover all active pharmaceutical ingredients (APIs) and all excipients used in the manufacture of pharmaceuticals.

Standard for API GMP will be the guide developed by ICH EWG Q7A (in preparation). In the field of bulk pharmaceutical excipients (BPEs), there is a broad consensus that the IPEC GMP Guide (1997) represents the state of the art and will serve as a guide for inspections of BPE manufacturers.

Continued from the previous page.

Chapter	Topic	PIC GMP (Integrated Chapters)	ISO 9001 (Integrated Elements)
9	Complaints and product recall	Complaints and product recall	Control of nonconforming products Corrective and preventive actions Product identification and traceability
10	Self-inspection	Quality control	Internal quality audits
11	Purchasing		Control of customer-supplied products Purchasing
12	Research and development		Design control
13	Safety and environment		Environmental management system (ISO 14001)
14	Quality controlling		Financial considerations of quality (ISO 9004) Servicing (not applicable)

Table 5.4. Amendment of an ISO 9001 quality system according to APIC (1998).

Chapter	Topic
0	Introduction
1	Management responsibility
2	Quality system
3	Contract review
4	Design and development control
5	Document control
6	Purchasing
7	Control of customer-supplied products
8	Product and equipment identification and traceability
9	Manufacturing process control
10	Inspection and testing
11	Control of inspection, measuring and test equipment
12	Inspection and test status
13	Control of nonconforming products
14	Corrective and preventive actions
15	Handling, storage, packaging, preservation and delivery
16	Control of quality records
17	Internal quality audits
18	Training
19	Servicing
20	Statistical techniques
Annex A	Hygiene
Annex B	Facilities and cleaning, utilities and engineering
Annex C	Validation
Annex D	Change control

The International Pharmaceutical Excipients Council (IPEC) favors another approach. They recommend integrating the GMP topics into the elements required by ISO 9001. In this case, Chapter 9, "Process Control", should be extended and completely rearranged (see Table 5.5).

Table 5.5. Structure of the chapter "Process Control" of an ISO 9002 system as proposed by IPEC (1997).

Chapter	Topic
9	Process Control
9.1	Buildings and facilities
9.2	Equipment
9.3	Water systems/water quality
9.4	Aseptic and sterile manufacturing
9.5	Validation of process and control procedures
9.6	Stability
9.7	Expiration/Reevaluation dating
9.8	Process changes (change control)
9.9	Lot/batch production records
9.10	In-process blending/mixing
9.11	Recovery of solvents, mother liquors and second crops
9.12	Impurities

QUALITY SYSTEM WITHOUT ISO 9001: THE MASTER PLAN CONCEPT

An approach to generating a documented pharmaceutical quality system in a non-GMP environment and an alternative to the integrated quality systems described above is the Master Plan Concept. In this approach, all GMP requirements are set forth in Master Plans. This concept can be applied to firms with or without an existing quality system.

Master Files

If one applies for the registration of a drug or a drug substance at the authorities in Europe or the United States one is forced to submit a Master File (EC 1998; FDA 1989; PIC 1993). Table 5.6 lists different types of Master Files.

A minimum requirement is the submission of a chemical pharmaceutical documentation (U.S. DMF Type II or EDMF). In

Table 5.6. Types of Master Files.

European Master Files		U.S. Drug Master Files	
PIC	Site Master File	Type I	Manufacturing site, facilities, operating procedures and personnel
EDMF	Chemical, pharmaceutical and biological documentation	Type II	Drug substance, drug substance intermediate, and material used in their preparation or drug product
		Type III	Packaging material
		Type IV	Excipient, colorant, flavor, essence or material used in their preparation
		Type V	FDA-accepted reference information

Europe there is another way to get a drug substance registered: if the substance is listed in the *European Pharmacopoeia*, the manufacturer can apply for a Certificate of Suitability to the Monographs of the European Pharmacopoeia (CEP) to the European Department for the Quality of Medicines (Council of Europe 1998). All these applications contain all documents for the production and testing of the product, as well as some more information specific for this product. In Table 5.7 the contents of a U.S. DMF Type II and a CEP application are listed.

In most cases it makes sense to generate a Site Master File which describes the firm, especially its organizational structure, premises and equipment, and its most important procedures. In Table 5.8, the structure of a Site Master File is shown. The International Inspection Convention (PIC) Document PH 4/93 gives detailed guidance on the preparation of a Site Master File (SMF). With these two documents, some of the basic requirements of a pharmaceutical quality system are met and documented. The remaining topics can be covered by Master Plans.

Master Plans

The hot topics of GMP, including validation and hygiene, as well as special requirements to certain topics, can be documented in

Table 5.7. Content of U.S. DMF Type II and the application for a CEP.

US-DMF Type II (example)	CEP (Application Dossier)
A Introduction	1 General information
B Composition	2 Chemical and pharmaceutical information
C Methods of manufacture and packaging	Manufacturing method
In-process controls	Quality control during manufacture
D Specifications and test methods	Development chemistry Impurities Batch analysis Technical characteristics
E Stability	Stability
F Environmental assessment	
	Expert report
	Potential toxicity of impurities

Table 5.8. Typical structure of a Site Master File.

Chapter	Topic
1	General information
2	Personnel
3	Premises and equipment
4	Documentation
5	Production
6	Quality control
7	Work contracted out
8	Distribution, complaints, product recall
9	Self-inspection

Master Plans. These Master Plans show the products and areas with respect to the departments where these regulations are effective and the responsibilities for the activities.

Validation Master Plan

The Validation Master Plan is a document commonly used to describe the firm's validation policy. The authorities demand that this document also contain a schedule of all validation activities. Detailed requirements for Validation Master Plans are set forth in the PIC/S Recommendations PR 1/99-1 Recommendations on Validation Master Plan, Installation and Operational Qualification, Non-Sterile Process Validation, Cleaning Validation (1999). Furthermore, the FDA has developed a Draft Validation Documentation Inspection Guide which contains requirements for a Validation Master Plan (1993), but this guide never has received official status.

A Validation Master Plan should include every validation activity relating to critical technical operations, relevant to product or process controls. To avoid misunderstanding, it must be clearly said that the term *validation* presupposes qualification, since a correct validation requires qualified manufacturing and control equipment. Although validation should be performed prospectively, the Validation Master Plan has to cover concurrent and retrospective measures as well. The structure of a Validation Master Plan according to the PIC document is shown in Table 5.9.

The content of the particular chapters could be as follows. Chapter 1, "Introduction", describes the firm's philosophy, intentions, and approach regarding validation and the scope of the Validation Master Plan (products, departments, premises). In Chapter 2, "Organisational Structure of all Validation Activities", the personnel responsibilities for the complete program as well as for single activities are laid down.

In Chapter 3, "Plant/Process/Product Description", all parts of the plant as well as processes and products which are subject to validation activities should be explained. A suitable way to make the processes easily comprehensible is the use of basic flow diagrams. Chapter 4, "Specific Process Considerations", describes specific characteristics and requirements of the plant or process (e.g., sterile manufacturing).

Table 5.9. Structure of a validation master plan according to PIC/S PR 1/99–1 (1999).

Chapter	Subject
1	Introduction
2	Organisational structure of all validation activities
3	Plant/process/product description
4	Specific process considerations
5	List of products/processes/systems to be validated
6	Key acceptance criteria
7	Documentation format
8	Required SOPs
9	Planning & scheduling
10	Change control

In Chapter 5, "Products/Processes/Systems to be Validated", all critical items to be qualified or validated should be listed. It should be specified which approach (prospective, concurrent, or retrospective) and which extent (installation qualification, operational qualification, or process validation) is chosen for the particular activities. This list should include the validation of analytical methods and the revalidation program as well. A suitable way to handle this document is to present this list of products, processes, and systems to be validated in matrix format as an annex to the Validation Master Plan.

Chapter 6, "Key Acceptance Criteria", requires only a general statement; it is not necessary to set down particular parameters within the Validation Master Plan.

The requirements for and the format of validation documents, especially validation protocols, records and validation reports, are described in Chapter 7, "Documentation Format". Chapter 8 lists the required standard operating procedures (SOP) which are relevant to carry out a validation program. Table 5.10 lists some SOPs which may be required for validation activities.

The Validation Master Plan is a management tool, too. Therefore a schedule for the complete program as well as for single activities should be included in Chapter 9, "Planning and Scheduling".

Table 5.10. Required SOPs for validation.

Design and construction qualification
Calibration and qualification guideline for suppliers
Organization of qualification
Organization of validation
Risk analysis
Qualification protocol
Validation protocol
Qualification report
Validation report
Qualification of premises and HVAC
Qualification of water and steam systems
Qualification of compressed air and gas systems
Qualification of the supply of electrical energy
Validation of manufacturing processes
Cleaning validation
Validation of computerized systems
Validation of analytical methods
Change control
Deviation control

It makes sense to integrate this schedule in the matrix of the annex to the Validation Master Plan, as described for Chapter 5.

A very important item is the control of changes. In Chapter 10 the firm's change control procedure should be described. This procedure has to ensure that everyone who is affected by critical changes to materials, facilities, equipment or processes including analytical methods is informed, and that the appropriate measures can be performed to prepare for the effects which may be caused by the change.

An example of a Validation Master Plan is shown in Annex 1 of this chapter.

Maintenance Master Plan

The Maintenance Master Plan is a document which sets forth the firm's policy regarding preventative maintenance. It describes maintenance and calibration in general and fixes the repetition intervals for certain measures. Therefore, a Maintenance Master Plan is a convenient addition to the Validation Master Plan for maintaining the qualified status of premises and equipment, and thus the validated status of the processes. Furthermore, the documentation of maintenance should be regulated, for example, by individual maintenance plans which set forth the maintenance program for the different equipment, forms for recording single measures during the performance and the reporting of the complete measure in the equipment's log.

Hygiene Master Plan

The hottest GMP topic aside from validation is hygiene. Particular activities during manufacture of pharmaceutical products have to be performed in defined areas. In the literature, some requirements for the handling of certain products are laid down (e.g., EC 1991 with annexes). These guidelines mainly cover sterile products and their preliminary products, but they can give ideas for the handling of other products as well. In a Hygiene Master Plan, the firm's cleanroom classes are defined and all processes are assigned to these classes. The personnel and material flow, between and within the different classes, has to be regulated, and the conditions, microbiological and physical parameters have to be checked regularly. Furthermore, procedures have to be described becoming effective on out-of-specification (OOS) results exceeding the threshold values (warning and action limits) of the monitoring program.

The Hygiene Master Plan should determine general requirements for the cleaning and disinfection of premises and equipment. These should be described in detailed single SOPs for particular pieces of equipment. Two topics that also should be covered by the Hygiene Master Plan are pest control and sewage and refuse disposal. Table 5.11 lists some SOPs which may be required for a suitable hygiene management.

It is clear that a Hygiene Master Plan can lay down only the general requirements for hygiene in a company. Since all facilities are different, the application of the requirements to certain plants should be described in particular hygiene plans.

Table 5.11. Required SOPs for hygiene program.

Clothing and behavior of personnel in controlled areas
Flow and handling of materials in controlled areas
Physical controls
Microbiological controls
Handling of out-of-specification (OOS) results
Pest control
Sewage and refuse disposal
Cleaning procedures for garment
Cleaning procedures for all premises and equipment

Training Master Plan

An important resource is the qualification of the personnel. A Training Master Plan describes what qualifications the workers performing certain processes should have, and what training they must receive. This includes the responsible personnel in all levels. Training needs should be identified and an appropriate training program should be established. It makes sense to create a standard information program for new employees, to assure that they become familiar with basic processes within the company and acquire basic knowledge on GMP before they begin routine duties. Table 5.12 shows the basic questions which should be applied to a training program.

There should be at least one general GMP training session and one hygiene training session per year. Furthermore, all processes have to be adequately trained, and all SOPs and other documents have to

Table 5.12. Basic questions about a training program.

Why	should this training be performed?
Who	should receive this training?
By whom	should this training be conducted?
How often	should this training be performed or repeated?
How long	should this training be performed?
What	should be covered by this training?

be introduced before coming into force. Additionally, it must be guaranteed that the trainers are also adequately qualified.

As in the other Master Plans, the requirements for documentation are laid down. There has to be a written training program and written records documenting the performance of the training. These records should contain at least the date and subject of the training, as well as the names and signatures of the trainer and the participants. The effectiveness of the training should be evaluated in a suitable manner.

APPLICATIONS

The Master Plan Concept is a way to generate a pharmaceutical quality system in a company, whether the company is certified according to ISO 9000 series or not. Therefore, numerous applications are possible in the case of manufacturers of pharmaceuticals and their starting materials.

Starting materials for the production of pharmaceuticals are often obtained from large chemical plants, which usually produce "technical quality". The amount used in the pharmaceutical production is often far below 1 percent of the total production. In some cases there is no quality system installed; therefore, the Master Plan Concept is a suitable approach to incorporating GMP in a non-GMP environment.

Companies which already have a quality system according to ISO 9000 series can supplement their quality system by Master Plans to make it a pharmaceutical quality system.

In cases where a pharmaceutical manufacturer has no formal quality system in place, the Master Plan Concept may be a way to generate such a system.

Because the ISO 9000 series will be revised during the next years, and these revisions will contain major changes including the loss of the 20-element structure of the present ISO 9001 standard (ISO 1998), the Master Plan Concept may serve best as an intermediate solution prior to generating a quality assurance system according to the revised ISO 9001 standard.

ANNEX 1—VALIDATION MASTER PLAN (EXCERPT)

Example	Quality Management		Document No. Q 0001-01
Effective from 01.07.1999	Replaces Document No. new	Revision annually	Page 1 of 3
	Validation Master Plan		

Table of Contents
1 Introduction
2 Organisational Structure
3 Plant, Processes, and Products
4 Specific Process Considerations
5 Products, Processes, and Systems to be validated
6 Key Acceptance Criteria
7 Documentation Format
8 Required SOPs
9 Planning and Scheduling
10 Change Control
11 Attachment

Approval				
	Author	Engineering	Production	Quality Assurance
Date				
Signature				

Example	Quality Management	Document No. Q 0001-01
		Page 2 of 3
Validation Master Plan		

1   Introduction

This Validation Master Plan specifies the validation activities for all "pharmaceutical grade" products which are manufactured by Example.

1.1   Validation Philosophy

Example manufactures high-quality "pharmaceutical grade" products. To obtain these products, all production and testing steps are carried out according to written procedures which ensure that these steps lead to the expected results. Therefore all critical operations relevant to product or process controls have to be validated.

Validation is the means of ensuring and providing documentary evidence that processes within their specified parameters are capable of repeatedly and reliably producing a product of the specified (required) quality.

Validation may only be performed if all manufacturing and control equipment involved in this process is qualified.

Qualification exercises assure through appropriate performance tests and related documentation and records that premises, equipment, and ancilary systems have been commissioned correctly and that all future operations will be reliable and within prescribed or specified operating limits.

Although validation should be performed prospectively or at least concurrently it is allowed to perform retrospective validation for products which are already marketed based upon accumulated data.

1.2   Scope

This Validation Master Plan covers all "pharmaceuticla grade" products. Therefore all departments, premises, equipment, and personnel of Example which are involved in production and/or testing of these products are regulated by this Validation Master Plan.

2   Organisational Structure

The head of the manufacturing department has the overall responsibility for the correctness of this Validation Master Plan and the carrying out of all validation procedures. The heads of the respective departments are responsible for the single activities carried out by their unit.

Example	Quality Management	Document No. Q 0001-01
		Page 3 of 3

Validation Master Plan

Annex 1

Schedule   Table 1

Code	System to be Qualified	IQ/OQ	Date	PQ	Date	Remarks
1 Premises						
B100	Plant B	retro	06/99	no	–	
B101	Weighing Cabin	pro	08/99	pro	09/99	only calibration
B102	Filling Cabin	pro	08/99	pro	09/99	only calibration
2 Equipment						
A1	Scale	no	–	no	–	
A2	Scale	no	–	no	–	
B1	Storage Tank	retro	06/99	no	–	
B2	Storage Tank	retro	06/99	no	–	
B3	Intermediate Storage Vessel	pro	07/99	no	–	
F1	Sterile Filter	*	*	*	*	*part of nitrogen supply
F2	Sterile Filter	*	*	*	*	*part of nitrogen supply
F3	Filter Unit	pro	07/99	PV	–	
R1	Stirring Vessel	pro	07/99	PV	–	
R2	Stirring Vessel	pro	07/99	PV	–	
T1	Drying Chamber	pro	07/99	PV	–	

retro=retrospective; pro=prospective; no=not required; PV=during process validation.

REFERENCES

APIC. 1998. Active Pharmaceutical Ingredients Committee—A sector group of CEFIC, *Quality Management System for active pharmaceutical ingredients manufacturers—Integrating GMP into ISO 9001*. Brussels: European Chemical Industry Council.

Council of Europe. 1998. Council of Europe Public Health Committee *Certification of suitability to the monographs of the European Pharmacopoeia* (Resolution AP-CSP (98) 2).

EC. 1989. European Commission *Guide to good manufacturing practice for medicinal products* (III/2244/87, Rev. 3).

EC. 1991. European Commission Directive laying down the principles and guidelines of good manufacturing practice for medicinal products for human use (91/356/EEC).

EC. 1993. Verordnung des Rates vom 29.6.1993 über die freiwillige Beteiligung gewerblicher Unternehmen an einem Gemeinschaftssystem für das Umweltmanagement und die Umweltbetriebsprüfung (Eco Management and Audit Scheme; EEC/1836/93).

EC. 1998. The rules governing medicinal products in the European Union, Vol. 2b. *Notice to applicants, presentation and content of the dossier.* Luxembourg: Office for Official Publications of the European Communities.

EC. 1999. Draft Proposal for a European Parliament and Council Directive on good manufacturing practice for starting materials of medicinal products and inspection of manufacturers.

FDA. 1989. *Guideline for Drug Master Files.* Rockville, MD: Center for Drug Evaluation and Research.

FDA. 1993. *Validation documentation inspection guide.* Rockville, MD: Center for Drug Evaluation and Research.

FDA. 1998a. Code of Federal Regulations of the Food and Drug Administration 21 CFR Part 211 *Current good manufacturing practice for finished pharmaceuticals.* Washington DC: U.S. Department of Health and Human Services.

FDA. 1998b. *Guidance for industry: manufacturing, processing, or holding active pharmaceutical ingredients.* Rockville, MD: Center for Drug Evaluation and Research, Center for Biologics Evaluation and Research, Center for Veterinary Medicine.

ICH. in preparation. International Conference on Harmonization, Expert Working Group Q7A *Good manufacturing practice for active pharmaceutical ingredients.*

IPEC. 1997. *Good manufacturing practices guide for bulk pharmaceutical excipients.* Amsterdam: International Pharmaceutical Excipients Council.

ISO. 1994a. International Organization for Standardization, ISO 9001 *Quality systems—Model for quality assurance in design/development, production, installation and servicing.*

ISO. 1994b. International Organization for Standardization, ISO 9004-1 *Quality management and quality systems elements—Part 1: Guidelines.*

ISO. 1996. International Organization for Standardization, ISO 14001 *Environmental management systems, specification with guidance for use.*

ISO. 1998. Working drafts of ISO 9001:2000 and ISO 9004:2000.

Metzger, K., and O. Schmidt. 1997. Zwei Konzepte in einem Guß [Two concepts in a unified whole]. *Chem. Prod.* 4:46–49.

PIC. 1987. Pharmaceutical Inspection Convention Document PH 2/87 *Guidelines for the manufacture of active pharmaceutical ingredients.*

PIC. 1993. Pharmaceutical Inspection Convention Document PH 4/93 *Explanatory notes for industry on the preparation of a Site Master File to be part of the information required under Article 2 of the Pharmaceutical Inspection Convention.*

PIC/S. 1999. Pharmaceutical Inspection Convention Document PR 1/99-1 Recommendations on Validation Master Plan, installation and operational qualification, non-sterile process validation, cleaning validation.

Schmidt, O. 1997a. Process orientation in quality assurance Part I. *Pharm. Technol. Eur.* 9 (10):40–44.

Schmidt, O. 1997b. Process orientation in quality assurance Part II. *Pharm. Technol. Eur.* 9 (11):36–44.

6

A Quality Manual for a Multinational Pharmaceutical Company

David F. Sullivan
Hoechst Marion Roussel AG
Frankfurt, Germany

WHY HAVE GLOBAL STANDARDS?

Many multinational pharmaceutical companies today were formed by the merger of two or more smaller companies, which themselves were probably formed the same way. When such large companies merge to create one large global company (in our case three companies merged), the people working in departments such as Global Manufacturing and Quality Operations (QO) may be faced with three different organisational work processes (and possibly three different cultural approaches) to achieve similar ends. All these work processes may have had good success in the past, and it may be difficult for the employees from each original company to realise why work processes and standards should be changed.

It is also true that some companies decide independently of such historical reasons to change and improve their own standards for quality and productivity reasons.

A Global Quality Manual is a very beneficial way to provide guidance to all sites in the group on how the merged companies' manufacturing, production and quality functions should work together in the future, and to capitalise on the best practices of each preceding company.

A functioning Quality Manual in the organisation gives a common language to the teams re-engineering potentially diverse global management practices and procedures. Having a common Quality System in operation can support the intercompany shipment of goods and also facilitate the transfer of production between sites when necessary.

WHY GLOBAL STANDARDS?

- To make merging three companies and three different ways of doing things easier.
- To be able to give guidance to our sites.
- To help our compliance audits in the future.

Distributing one predefined manual to all concerned is not an ideal solution; rather, there should be a coming together of the best practices from each of the legacy companies into one overall quality philosophy and approach. This project requires an overall, top-down management approach, but this is beyond the scope of this chapter.

BACKGROUND

Typically, the pharmaceutical industry has relied on a very structured procedural/regulatory approach to quality, as opposed to the systems-driven Total Quality approach, as exemplified in the works and writings of W. Edwards Deming (1986) and in the International Organization for Standardization (ISO) 9000 series of quality standards. The existing quality systems in classical pharmaceutical companies have relied heavily on the stimulus and response method: the plane crashed because of factor X, so we fix X in all the other planes. In many individual factories, quality procedures have grown like weeds as a result of an individual problem. This may work well until the moment a new quality assurance (QA) supervisor is faced with 1,500 extant standard operating process (SOP) in his area of responsibility. Many of these are probably purely academic, as no one in the factory could have read *all* these SOPs and their revisions. There is almost certainly no overall theme or connection running through these SOPs, as they were all developed and written at different times by different people.

A risk exists of the same thing happening to a Global Quality Manual. For that reason, it is ideal if the operational management can agree on an overall process for preparing such a quality manual.

HOW IS A QUALITY MANUAL DEVELOPED?

The following sections outline an approach for developing a Quality Manual.

Step 1 involves the formation of a small working group of quality professionals who have wide-ranging experience and a good knowledge of the company, including each of its individual legacy components. Their objective is to prepare a proposal for the wider range of global quality management to buy into. This proposal will look into the "as is" status of the present company's systems and outline a "to be" approach to be followed in the future.

In Step 2, the quality management group, having received wider management approval to continue, needs to agree on the scope and structure of the proposed quality manual. This is an important step as it provides the "Quality of Design" for the total quality systems that will be fostered in the corporation for the next 5 to 10 merger-free years.

In the case of our multinational and multicultural company, the wider management group decided on the following scope and hierarchy.

SCOPE

Given the diverse nature of a multinational company trading in more than 100 countries with different standards of living, capacities, customer expectations and regulatory environments, standards could not match the highest one-country requirement. Neither could they be the lowest common denominator of standards. Instead, their development was based on the following concepts:

QUALITY STANDARDS

- Cover the "vague areas" of responsibilities.
- Are different from the lowest common denominator.
- Are different from the highest "one country" requirements.
- Are written in drafts and circulated to all partners.
- Are finally approved by the corporate quality department.

Quality policies and standards in a multinational pharmaceutical company are written to tell all employees what the company expects of everyone with respect to the quality of products and services. In many cases, they cover vague areas of responsibility, or areas where divergent practices may be in place for historical reasons or because of recently merged companies.

We agreed that it would serve no purpose to duplicate many of the existing global regulatory and industry standards that exist in all the countries. But at the same time our table of contents had to cover all of the well-defined topics of Good Manufacturing Practice (GMP). We settled on a manual with 16 sections comprising approximately 100 standards. By the completion of the project we arrived at 110 standards.

QUALITY STANDARDS WERE WRITTEN BASED ON

- Practical experience in the industry.
- Appropriate legal requirements in the strategic countries.
- Input from all sites.
- Corporate considerations.

Some of the standards by their nature are written from the top down and are not negotiable, but in most cases they are developed as a result of a wider democratic process.

The Quality Manual requires that the sites producing products for our company meet the GMP standards of the World Health Organisation (WHO), as well as local regulatory requirements in the markets where the products will be sold and our own internal company standards.

A basic decision was made also on the use of this Quality Manual by individual legal entities around the world. The manual as written is intended to be an internal company document and is to be used by the sites in the writing of their own site standards and SOPs. It is not intended to be used for any local legal purposes. The manual was written with the concepts of good document management practices and change control in mind but is not treated as a GMP document itself. The local sites are obliged to use the SOPs written in their local language for any official exchanges with their local regulatory authorities or for any business arrangements.

HIERARCHY

The hierarchical structure chosen for the Hoechst Marion Roussel Quality Manual (illustrated in Figure 6.1) is a straightforward structure divided into the two main management domains, one for the corporate global management and the other for the global sites. For the global management part we have the quality policies on the highest level. This is similar to the legal system in each country, with the actual text of the law being short and succinct. In the same way, our policy statements are for the most part one or two sentences. The one exception to this is the overall company quality policy, which is almost one page long.

Beneath the policies come quality standards, which are more detailed (1–3 pages) narratives than the policies, but are still written as relatively high-level documents (containing "what" to do but not necessarily "how" to do it). Following the legal analogy, these standards are similar to legal regulations.

On the lowest level are the more detailed global guidelines, which provide the "how" to do things, including detailed examples, recommendations and so on.

Figure 6.1. Quality system hierarchy.

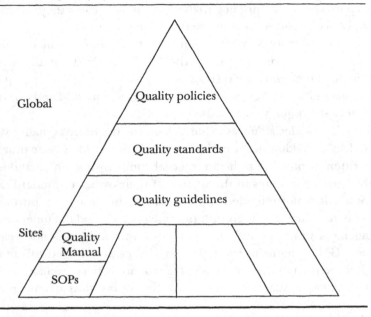

The individual sites are required to link their site SOPs to the individual quality standards and guidelines. In this way we can maintain a global structure and hierarchy, giving all the sites an anchoring point to which they may connect their existing quality structure. This also served very well to allow us to maintain an existing site quality system or SOP structure while allowing the site to evolve into the new structure. It is impossible to think that, for a process that was to take two years to implement worldwide, a site could (or would) scrap its existing structure and replace it with the new one all in one move.

PROCESS TO GENERATE STANDARDS

Part of the success in the implementation of the standards globally lies in the involvement of the global partners in the development of these standards. When people are consulted and are part of a development process, then at the implementation stage there will be more understanding of the background behind the individual standards, and therefore a faster implementation. We found that the response to the drafts and the ensuing debates were very valuable for all of the process partners.

Input to generate the standards can come from many sources, including the management, the customer requirements, the global sites, specifically formed subgroups of experts and so forth (see Figure 6.2). The input thus gathered can subsequently be used to form a draft and become part of the first expert review, as discussed later.

In our case, the final review and approval of the drafts was provided by a multifunctional management team with representatives from other global functional management groups such as Manufacturing Operations, Development, Regulatory and so on. All these types of measures help to communicate the process of developing a Quality Manual to a wide audience within the company, and allow full participation and input from all the global partners.

Gathering input from the global sites means having to slow down the process of completing the standards. Drafts have to be sent out to site contacts who in turn must solicit input from their site colleagues and then discuss and consolidate the comments back up through the management line. Connecting the sites via an electronic mail system can reduce the time spent in this process.

We realised that having draft standards in good condition prior to general distribution would mean a faster and better quality review. Badly written draft standards risk the problem of confusing

Figure 6.2. Process used to generate standards.

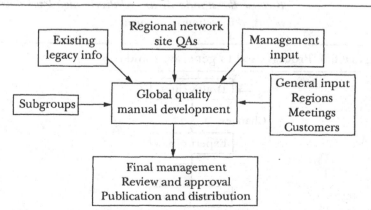

people and increasing the cycle time as they may be rewritten and sent out again.

To increase our probability of having a "good" draft, we decided to have an expert write the draft standard, and then circulated the first draft to a small group of selected experts for review and refinement prior to general distribution (see Figure 6.3). This first review process was typically fast and showed that we could reduce the confusion and debates by sending a draft that was 80 to 90 percent of the expected final version out for global review.

STANDARDS CONTENT

The major part of our Quality Manual is the quality standards, as they show the individual policy statements and the corresponding scope, responsibilities, and requirements associated with the policy. Ideally these standards are high-level documents outlining "what" has to be done in as few words as possible.

Each standard was composed of the following five sections.

Policy: Briefly states the company's requirements or objectives.

Scope: Outlines the field of application of the standard.

Responsibility: Describes which management groups or functions are responsible for what.

Because the standards are relatively high-level documents, this section is important.

Figure 6.3. Process used to generate standards.

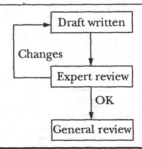

Requirements/   Outlines the requirements of the individual stan-
Process: dards, ideally including a process flow diagram.

History: Maintains an historical record of any changes that
                are made to the content.

IMPLEMENTATION STRATEGY

Ownership

Having the Quality Manual written and available, of course, is only part of the overall story. The bigger challenge is worldwide implementation.

We decided early on that the implementation of the manual would be done by the responsible line management and not the corporate group. In our case, this meant the regional management teams. These are the people running the organisational and geographic units. They are the people that can balance the rate of implementation, based on individual site strategy and operational imperatives. For that reason, we decided that the prioritisation and the speed of implementation would be controlled by the regional management teams within an overall global time frame. With the time frame for the revision of the individual site SOPs being two to three years, this would be the fastest way to incorporate the global Quality Manual requirements into local SOPs. As our experience later showed, some regions and sites were better equipped to implement these standards faster than others.

This mechanism allowed us to have the standards in force on a global basis without having to wait until all sites were in a position to implement the standards.

IMPLEMENTATION

- Standards are effective on date of publication.
- Regional quality department is responsible for implementation schedule.
- Local sites are responsible for local implementation.

Roll-Out

If the standards are seen as something abstract coming from the corporate group, their chances of successful implementation will be low. In roll-out of the standards, we stressed that ownership of the standards was the responsibility of all the line management. It is well known that, when many people are sharing resources, unless some one feels responsible no one will take responsibility.

GLOBAL QUALITY STANDARDS

> • Who are the owners of the standards?
>
> • If you are responsible for any GMP-related item in a site, then *you own* the standard. You are responsible for compliance in your area.

Gap Analysis

Each organisational region set up a "Gap Analysis" process in their sites which allowed the regions to map the potential work load involved in bringing all the sites in line with the standards. This process asked the sites to go through the standards one by one with the local management groups and make detailed assessments on the status of the sites. This then allowed the regional management teams to develop programs to work with the sites in terms of effort, training or investment capital required at each site.

AVAILABILITY

We perceived availability as a simple thing, but it proved to be more difficult than expected. Making available more than 400 copies of a document containing more than 300 pages while maintaining good *change control* was not easy. To make the manual readily available for all the personnel in the global sites, we had wanted to put the whole document on the Internet only (no paper copies to distribute). However, the installation of the Internet backbone was running in parallel with the development of the global standards, and we decided that we could not risk the chance of some colleagues not being able to access the Quality Manual because of unrelated information systems reasons. For that reason we decided to publish the manual in hard copy and on the Internet.

VALUE ADDED

For companies to invest in these type of efforts there must be return on investment. The Quality Manual was developed with the clear understanding that it should add value and not only add costs.

Having diverse technical groups based in over 70 worldwide sites using a common technical language (nomenclature) and a common understanding of the overall company requirements is a huge advantage for a multinational company.

QUALITY STANDARDS

VALUE ADDED FOR SITES

- Platform to discuss common issues and common standards with colleagues.
- Allows formalised system for communications on quality issues.

VALUE ADDED FOR CORPORATION

- Reduces risk of surprises.
- Helps to achieve compliance at a regional and global level.
- Creates common global quality systems.

It is hard to measure the sometimes intangible cost savings associated with a published Quality Manual in a large corporation. Our experience since the standards were introduced has shown that having all the diverse sites speaking in a similar language about similar problems has been a huge benefit. At the beginning of the process, simple discussions among the sites were almost impossible to resolve because of the different understanding of the words and concepts being used by the colleagues. Today, with most of the major company quality-related business process written down, understood, and used in conjunction with a company glossary, management discussions can move on easily to higher levels.

A consistent approach to quality systems in the global sites means more transparency for the regulatory authorities and thereby increases the flexibility for the corporation in terms of sourcing changes. More standardised approaches in the company's quality systems translate into faster cycle times in new product introduction, as well as better control and handling of potentially adverse events such as complaints and recalls.

The job of the corporate auditors is also greatly facilitated by having a clear understanding of stated company requirements at the site level.

Similarly, for the individual sites the opportunity to participate in the discussion phase during the development of the standards helped increase their levels of understanding and fostered individual discussions between the sites.

For the management groups responsible for the global sites, the process and individual debates around each of the standards helped to give them a better comprehension of the quality systems and the compliance levels in the sites under their jurisdiction.

Company standards are never static, and the process of updating and improving will be a continuous process.

REFERENCES

Deming, W. Edwards. 1986. *Out of the crisis.* Cambridge, MA: MIT. ISBN 0-911379-01-0.

Scherkenbach, William W. 1987. *The Deming route to quality and productivity.* Mercury Press. ISBN 0941893-00-6.

Quality Systems—Model for Quality Assurance in design/development, production, installation and servicing. ISO 9001.

7

Implementation of a Quality Assurance System into a German Pharmaceutical Company

Heinrich Prinz
Biotest AG
Dreich, Germany

Quality Assurance (QA) Systems (sometimes also called "Quality Management" in German laws, European Economic Community (EEC), European Commission (EC) and European Union (EU) and other international directives, guidelines and guidances are playing an increasingly important role in the pharmaceutical industry. Legal and other prescribed conditions which should be observed, such as EC Good Manufacturing Practice (GMP) directives (EC-GMP Directive 1991), Operation Ordinance for Pharmaceutical Entrepreneurs PharmBetrV (Betriebsverordnung für Pharmazeutische Unternehmer 1994), International Inspection Convention (PIC) Document PH 1/97, World Health Organization (WHO) GMP (WHO 1992) and CFR 21 parts 200 to 299 (CFR 21 1998) demand a QA System that ensures the medicinal product is of the required quality for its intended use.

The active involvement of the top management and the personnel performing the tasks is also required.

In addition to filling the legal requirements, companies are also obliged to take account of and satisfy further factors relating to the prescribed conditions. As none of the framework conditions provide a specific definition of a QA System or its method of function, and as they do not present any further insight into its detailed implementation within a company, assistance can be taken from other non-branch-specific regulations such as the German Institute for Standardization (DIN) EN ISO 9001/4 DIN EN ISO 9001, 1994; (DIN ISO 9004, 1987).

Systems that rationally and reasonably satisfy the comprehensive legal requirements of a QA System can only be established by a completely new structure within the company. Furthermore, the initial implementation, continuous monitoring and further extension of this system can be rationally and economically achieved only by personnel or departments dealing exclusively with the legal requirements for a QA System, with the help of the responsible persons according to European and German law.

The best method of building up a specific company's QA System is based on analysis of the current situation. This can be done by an inspection or auditing inside the company in every department. During this inspection, conformity with the regulations chosen for this department can be checked and deviation can be mentioned and written in a report.

Based on an analysis of the current situation, a specific QA System can be built up in accordance with legal requirements such as national laws, GMP Directives, GMP Guidelines and internal requirements. It can then be enlarged by implementing relevant elements from the international standards DIN EN ISO 9001/4.

Due to the national laws that German companies must follow, they must first implement a QA System according to Pharm-BetrV and EC-GMP Directive/Guideline. This first step in the building and implementation of a QA System can be announced as a basic system. The second step deals with its extension according to DIN EN ISO 9001/4, and including relevant elements from this international standard, such as the responsibility of the management or the design of new or changed products.

A QA System that complies with both mentioned regulations can then be termed a complete QA System.

In this chapter, primary attention is given to German national laws and the relevant European regulations which must be followed.

Some references are made to WHO guidelines and Food and Drug Administration (FDA) requirements to show the similarities and sometimes the differences between German and U.S. regulations.

Biotest AG as a pharmaceutical company began the 1990s with the implementation of a new department called *Zentrale Qualitätssicherung* (QA as a department linked directly to the top management). Its responsibility is to implement, enhance and support the QA System for the entire company. That means the personnel of this department have to give a helping hand in the interpretation of new or existing regulations, translate these into the daily work fixed by standard operating procedures (SOPs) and observe and audit these internal and external regulations. It is also the work of the department to train personnel in the relevant regulations (internal SOPs, laws, directives, guidelines and so on).

This department is directly linked to the top management— in Biotest AG, the board of directors—and reports to them.

THE QA SYSTEM—A CURRENT VIEWPOINT ON NATIONAL AND INTERNATIONAL LAWS AND REGULATIONS

The QA System for pharmaceutical companies in Germany is a legal requirement of the PharmBetrV and, at the EC level, of the directive 91/356/EEC (EWG Richtlinie 1991), the EC GMP directive, and therefore the EC GMP guideline (EG GMP Leitfaden 1998) which are both implemented in national law.

The German Medicines Act (*Gesetz über den Verkehr mit Arzneimittel*, AMG) (AMG 1998) does not stipulate any regulations for the QA System in its current edition.

Starting with the international legal demand based on European law, Chapter II, Article 6 of the EC GMP directive stipulates the following requirement for a QA System:

The manufacturer shall establish and implement an effective pharmaceutical Quality Assurance System, involving the active participation of the management and personnel of the different services involved.

EC GMP directive, Chapter II, Article 6

The EC-GMP guideline for medicinal products relevant to this directive defines the system as follows:

> Quality Assurance is a wide-ranging concept which covers all matters which individually or collectively influence the quality of a product. It is the total sum of all organised arrangements made with the object of ensuring that medicinal products are of the quality required for their intended use. Quality Assurance therefore incorporates Good Manufacturing Practice plus other factors outside the scope of this Guideline.
>
> *EC GMP Guideline, Chapter 1, Quality Management*

The PharmBetrV, the regulatory organ for ensuring the proper manufacture of medicinal products in the area regulated by the AMG, describes the system as follows:

> Businesses and institutions must implement a functional pharmaceutical Quality Assurance System according to the type and extent of their activities being performed in order to ensure that the medicinal product is of the quality required for its intended use. This Quality Assurance System must involve the active participation of the management and personnel of the services concerned.
>
> *PharmBetrV, §1a, Quality Assurance System*

If a system of this type is to be built up in a pharmaceutical company, the awareness of its composition and an accurate definition of the system are important. Several articles have already attempted to give a basic understanding of it (Völler 1997; Mühlen 1992; Häusler 1993). However, they are of little benefit for brand new construction. What is meant by a "QA System"? How is it structured? How can it be constructed within and for a specific company? And how can its function be demonstrated?

The GMP Guideline describes the QA System by the establishment of individual tasks and their relevant documentation. The individual tasks are grouped into three major blocks:

- QA in general, subdivided into points I to IX;

- GMP for Medicinal Products, subdivided into points I to X; and

- Quality Control (QC), subdivided into points I to VIII.

The previous mentioned individual blocks 1.3 and 1.4 describe specific requirements for the production and quality control departments in form of documented items to show evidence of a Quality Assurance System inside. This is a part of the general—an overall—Quality Assurance System.

It can be seen that Good Manufacturing Practice and Quality Control are linked to each other to form a solid cornerstorne in the basic concept.

Block 1.2 gives evidence to build up a system of Quality Assurance designed as a structure or basic task inside the daily work.

Therefore, the documentation of manufacturing and quality control, together with a Quality Assurance Department watching the documents and the structure, give evidence of a basic quality assurance system.

It can be seen that GMP and QC are linked to each other and form a solid cornerstone in the basic concept of a QA System. In addition, it is mentioned in Chapter 1 of this guideline that there may be factors outside the scope of the guideline which should be part of the system.

Nevertheless, these detailed explanations are of little benefit for a company planning to implement a QA System.

It describes only which points must be followed to assure that a pharmaceutical product will be of appropriate quality. Unfortunately, the PharmBetrV also gives no further explanations or instructions as to how this should be achieved.

WHO's good manufacturing practice for pharmaceutical products defines a QA System in a way similar to the European directive and guideline.

In this definition, QA is described as a concept rather than a system. Quality systems are covered in Part One, dealing with quality management in the drug industry.

The QA described in WHO's GMP refers to a system subdivided into points (a) through (i).

Quality Assurance is a wide-ranging concept covering all matters that individually or collectively influence the quality of a product. It is the totality of the arrangements made with the object of ensuring that pharmaceutical products are of the quality required for their intended use. Quality Assurance therefore incorporates GMP and other factors, including those outside the scope of this guideline such as product design and development.

WHO's GMP, Annex 1, Good manufacturing practices for pharmaceutical products

In the drug industry at large, quality management is defined as the aspect of management function that determines and implements the "quality policy", i.e., the overall intention and direction of an organization regarding quality, as formally expressed and authorized by top management.

The basic elements of quality management are:

- an appropriate infrastructure or "quality system", encompassing the organizational structure, procedures, processes, and resources; and

- systematic action necessary to ensure adequate confidence that a product or service) will satisfy given requirements for quality. The totality of these actions is termed "quality assurance".

Within an organization, quality assurance serves as a management tool. In contractual situations, quality assurance also serves to generate confidence in the supplier.

WHO's GMP, Annex 1, Good manufacturing practices for pharmaceutical products. Part One. Quality management in the drug

In opposition to the European and German understanding of a QA System defined according to WHO's GMP, Production and QC *together with* other points such as complaints, validation, product recall and contract production and analysis are part of a quality management philosophy and its essential requirements.

What *exactly* is a *QA System?* How can it be defined to facilitate implementation into a German pharmaceutical company?

DEFINITION OF THE TERM *QUALITY ASSURANCE SYSTEM*

The regulations cited do not provide a clear and understandable definition for a QA System. These regulations only describe what a QA System must ensure. It has to ensure that every step in the manufacturing of a pharmaceutical product is well implemented and documented together with GMP for the Production and QC departments.

To give a clear definition, the term should be broken into two parts: quality and system.

What can be understood by *the system* that must be implemented?

The concept of a system originates from the Greek and is defined as "anything formed of parts placed together; a set of data, procedures, and things considered connected as an entirety".

Thus, systems do not exist on their own but rather require support in order to ensure the necessary connection between data, records, documents and the relevant procedures. A system must therefore be built up by persons from individual departments of the company so that the individual departments can then interact and function as an entire system.

What can be understood by the concept *quality?*

According to AMG § 4 15, quality is defined as follows:

> Quality is the property of a medicinal product that is determined according to identity, assay, purity and also other chemical, physical and biological properties or by the manufacturing process.
>
> *AMG, §4, (15)*

This means the highest goal of a pharmaceutical company is to produce a product of the highest quality and the lowest risk for patients.

The system must therefore assure that the quality of the daily work produces the necessary product quality.

With this in mind, a new and more precise definition of a QA System can be offered:

> A Quality Assurance System plans, describes, and documents every step of the manufacture and the essential environmental require-ments so that the required pharmaceutical quality of the products is produced, controlled, and maintained at every level with the help and observance of the internal and external regulations. It must be supported and extended by responsible persons.

THE THREE COLUMNS OF A QA SYSTEM

If the procedures concerning the manufacture of a medicinal product in a pharmaceutical company are considered in detail, a distinction can be drawn between three individual systems. They must each be well established to guarantee a highly qualified product. As interconnected individual systems, they can ensure the requirements of a QA System and define and support the entire system throughout the company (see Table 7.1).

Each individual system described here can exist indepen-dently from other departments in its own function area. However, each must be continually revised, supported and developed.

Each individual system, however, gives only very partial sup-port to the comprehensive safety of medicinal products, starting from the procurement of the starting materials up to the safe use in patients. It is very difficult for individual systems of this type to create a thorough and functional QA System throughout the entire company.

System Supporting the Process Quality

As Table 7.1 demonstrates, the "system supporting the process quality" must ensure that the medicinal product is manufactured according to the registered and validated process.

According to the national law, most of the pharmaceutical products such as medicine have to be registered by national author-ities prior to selling them on the market. The basis of this act is

---

Table 7.1. Definition of the QA System. The pharmaceutical QA System is built from three individual systems, which together form three columns that support the complete system.

The QA System must:

- Assure that the medicinal product is manufactured according to registered and validated processes;

Process Quality

- Prove that the medicinal product satisfies the specified requirements; and

Product Quality

- Assure that the framework conditions for the personnel, the documentation, and the organisation are present and sufficient.

Infrastructural Quality

---

the proof of documents describing, for example, the process of production, the methods running during in the laboratory, the validations and the biological and medical trials on the new product.

During this registration stage, a final file concerning the mentioned documents is signed as a "contract" on both sites and the manufacturing process must follow the requirements set forth therein.

Therefore GMP is obliged to pay attention to the internal SOPs as well. Changes in either the manufacturing process or the corresponding SOPs must be controlled very carefully and sometimes discussed with the authorities if these can be outside the contract (see Table 7.2).

System Supporting the Product Quality

The system supporting the product quality must demonstrate that the medicinal products satisfy the specific requirements. It must also assure that the product quality, as defined according to AMG, demonstrates the lowest possible risk to patients.

This must be done by testing products according to specified requirements with validated and predefined procedures.

Table 7.2. Examples that determine the process quality.

- Established and documented operating procedures.

- Reproducibility of the manufacturing process.

- Hygienic harmlessness of the surrounding area.

- High-quality ingredients.

- Validated manufacturer process.

- Product and process according to registered documentation.

- Well-maintained equipment.

As discussed previously, observation of the relevant parts of the European GMP regulations is also necessary (see Table 7.3).

System Supporting the Infrastructure Quality

Companies manufacturing pharmaceutical products must follow the European GMP regulations as they are adopted into national law; in Germany this was done through the "Betriebsverordnung für Pharmazeutische Unternehmer", a regulation governing the pharmaceutical companies. The interpretations of a QA System are nearly the same in both regulations, but the ECGMP guideline gives more detailed information on the items to be implemented.

Table 7.3. Examples that determine the product quality.

- Examination of the quality of the product; in-process and end-process control.

- Examination of harmlessness.

- Examination of efficacy.

- Complete documentation.

- Validation and release procedure.

- Meeting of specified requirements.

- Retention of samples to demonstrate the correctness of the production on request.

One can therefore assume that transferring the points in Chapter 1 and preparing the documentation demanded in Chapter 4 for the manufacturing process and in Chapter 6 for QC should be sufficient to demonstrate a documented QA System.

The national authorities in Germany have to inspect the pharmaceutical companies (§64 AMG) and their QA Systems mainly with respect to product and process quality systems.

It becomes more and more obvious that the points mentioned in Table 7.4, together with the previous points, are important facets of a functioning QA System supported by the top management and the responsible person(s) (see Table 7.5).

The WHO GMP directive gives some more detailed information on how to build up a system that supports the infrastructure quality by the responsibility of the top management. It divides the QA System, or as it is called here, the *Quality Management System*, into basic elements, including essential requirements for the top management.

Table 7.4. Examples of infrastructural quality supporting the QA System, besides the GMP-linked Support of Process and QC.

1. Organisation charts.

2. Actual job descriptions not only for qualified persons).

3. Continuous training of staff and management.

4. System of audits or self-inspections, and corrective and preventive actions taken therefrom.

5. Clearly defined responsibilities expressed in job descriptions or in a responsibility matrix released by the top management.

6. Discussion of the audit reports with the heads of departments.

7. System for working intensively on internal or external complaints.

8. Control of corrective and preventive action—based complaints by the top management.

9. Qualified and sufficient personnel in every department.

10. Management of interfaces by the top management.

Table 7.5. Examples that determine the infrastructure quality supported by the top management.

- Defined organisation.
- Defined responsibility.
- Specified competencies.
- Management of interfaces.
- Written prescribed conditions and instructions.
- Qualified trained personnel.

WHO GMP The basic elements of quality management are:

> an appropriate infrastructure or "quality system", encompassing the organization structure, procedure, processes and resources: and systematic actions necessary to ensure adequate confidence that a product or service) will satisfy given requirements for quality. The totality of these actions is termed "quality assurance".

It should be noted here that the WHO GMP guidelines relating to a QA System refer to the DIN EN ISO 9001 standards as a source.

In summary, on the basis of what has been discussed, a Pharmaceutical QA System can be regarded as a composed system supported and built up by the main three columns of system, as described in Figure 7.1.

SUPPORT OF THE THREE COLUMNS BY DEFINED RESPONSIBILITIES ACCORDING TO GERMAN LAW

Who of the responsible persons, according to legally prescribed conditions, should primarily ensure the support of the individual systems listed in Table 7.6?

Figure 7.1. The pharmaceutical QA system, composed of regulations and the three columns of quality.

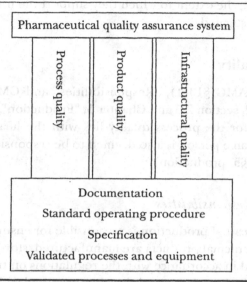

The GMP Guideline (Chapter 2, Personnel, Subsection: Key Personnel) nominates the head of Production and also the head of QC, on the basis of EEC directive 75/319/EEC on detailed responsibilities in subsectors 2.5 and 2.6, respectively. Both are thus responsible for supporting the QA System with regard to process quality and product quality, as will be shown.

In the AMG (§ 19, Responsibilities), the head of QC, the head of Production, the head of Sales, the person in charge of information and the person in charge of complaints in hospitals (Stufenplanbeauftragter) are nominated to share the responsibility in different fields.

In order to ensure the proper performance of procedures and tasks allocated to their job, the job descriptions are intended for persons in management or other responsible positions. Furthermore, the organisational structure should be described in an organisation chart. The persons nominated must be given sufficient authority to be able to execute their responsibilities.

As the competences in the areas of responsibility are named according to the AMG (and stipulated in the job descriptions) and exactly defined, the extent to which they support the total system should be investigated.

Process Quality

According to AMG §19(1), "Responsibilities" and GMP Guideline, Chapter 2, section 2.5 and Chapter 5, "Production", personal responsibility for the process quality lies with the head of Production. The same person is also deemed to be responsible by the PharmBetrV (§5, production).

AMG: §19 Responsibilities

1. The head of production is responsible for ensuring that the medicinal products are manufactured, stored and labeled in accordance with the regulations on trade with medicinal products, and that they are fitted with the specified package inserts.

GMP Guideline: Chapter 2, "Personnel" and Chapter 5, "Production"

2.5, I The head of Production must ensure that the products are produced and stored according to the appropriate documentation in order to obtain the required quality.

5.1 The process should be performed and supervised by competent staff.

PharmBetrV: §5, Production

1. Medicinal products are to be manufactured according to acknowledged pharmaceutical rules.

2. Medicinal products are to be manufactured and stored under the responsibility of the head of Production and according to previously drafted instructions and operating procedures manufacturing specifications).

It is important to mention here that manufacture and testing must always be in accordance with the valid registration documents. According to PharmBetrV §5 and §6, the responsibility for this lies with the head of Production and/or the head of QC. According to Chapter 5 of the GMP guideline, it lies with the Head of Production and should be confirmed in documented form by both persons.

Product Quality

According to AMG §19(2), "Responsibilities", PharmBetrV§6(1) and (2) and GMP guideline, Chapter 2, Point 2.6, in conjunction with Chapter 6, "Principle", the QC Department and thus the head of QC (as the persons designated responsible) are responsible for testing and the quality of the medicinal product.

AMG: §19 Responsibilities

2. The head of QC is responsible for ensuring that the medicinal products are tested for the necessary quality according to the regulations on trade with medicinal products.

PharmBetrV: §6 Testing

1. Medicinal products and the starting materials are to be tested for the required quality according to acknowledged pharmaceutical rules.

2. The tests are to be performed under the responsibility of the head of QC and according to previously drafted instructions and operating procedures test specifications).

GMP Guideline: Chapter 2, "Personnel" and Chapter 6, "Quality Control"

2.6, III   The head of QC must ensure that all necessary testing is carried out.

Chapter 6:   The quality control . . . ensures that the necessary and relevant tests are carried out and that materials are not released for use, nor products released for sale or supply until their quality has been judged satisfactory.

Infrastructural Quality (System Quality)

A defined responsibility for a general infrastructure quality can be derived from AMG §13 in conjunction with §14, in which a manufacturer's license is required. This license can be revoked only if the individually required and qualified persons are not nominated (points 1–5) or if suitable rooms and equipment are not available (point 6).

Similarly, a general responsibility is also expressed in Pharm-BetrV §1a and GMP Directive Chapter II, Article 6, in which the active participation of the top management is required without explanation of any practical reason.

Specific parts of infrastructure quality can further be covered by cooperation between the head of Production and the head of QC.

Basic tasks are defined by observing and following the breakdown of areas of responsibilities according to Chapter 2, Section 2.7 of the EC GMP Guideline. The head of Production and the head of QC share or jointly exercise responsibilities relating to quality. These may include, subject to any national regulations:

- the authorisation of written procedures and other documents, including amendments;

- the monitoring and control of the manufacturing environment;

- plant hygiene;

- process validation;

- training;

- the approval and monitoring of suppliers of materials;

- the approval and monitoring of contract manufacturers;

- the designation and monitoring of storage conditions for materials and products;

- the retention of records;

- the monitoring of compliance with the requirements of GMP; and

- the inspection, investigation, and taking of samples in order to monitor factors which may affect product quality.

Responsibility for the Infrastructural Quality

Objective consideration of the points listed reveals that the three individual systems (columns) of a QA System can only be supported by the head of Production and the head of QC for their own department they rule.

But the combination of all three columns into a complete company-wide QA System is rarely achieved by them and can be done correctly only by the head of each department, only for *his or her own* department. (This includes such tasks as training, archiving of records, approval, monitoring of suppliers and so on.)

A company-wide coordination is rarely achieved, and is especially difficult if the company has reached such a size that different departments in different fields and services are required to cooperate to perform individual tasks, that is, when tasks cross departmental or service group boundaries.

The manufacture, testing, sales and the assurance of all peripheral conditions during production of commercial products, however, can never be done in isolation by individual departments or by individual responsible persons listed above. It must always be coordinated. It is further necessary to agree upon, implement and monitor common measures and tasks across departmental and service boundaries.

In order to ensure the rational cooperation of all relevant departments and nominated responsible persons, and thus also to ensure the functional ability of the company, all the aforementioned procedures discussed as well as others not specified here must be combined to form a single structure that guarantees a seamless interlocking of the tasks listed (managing interfaces).

The GMP guideline expresses the aim and responsibility of a QA System more clearly than does the PharmBetrV.

In contrast to the EC GMP Guideline, the PharmBetrV states only the requirement that the management be involved in an active manner. But it does not clearly define or explain how management should go about its involvement.

In the guideline we can find the information that the management is responsible for the implemented QA System. To be responsible is already something more than being involved.

Implementation of the Infrastructural Quality

Many of the necessary responsibilities for the infrastructure quality can be covered and met by the top management (organigram, job description, stipulation of responsibilities and competences). However, there are certain tasks these people can perform only with very great difficulty, and thus should be subject to a different type of monitoring and common regulation interfaces, validations, trained personnel, system auditing and so on.

The holder of a Manufacturing Authorisation must manufacture medicinal products so as to ensure that they are fit for their intended use, comply with the requirements of the Marketing Authorisation and do not place patients at risk due to inadequate safety, quality or efficacy. The attainment of this quality objective is the responsibility of top management and requires the participation and commitment by staff in many different departments. . . .

To achieve the quality objective reliably, there must be a comprehensively designed and correctly implemented system of Quality Assurance incorporating Good Manufacturing Practice and thus Quality Control.

This system should be fully documented and its effectiveness monitored.

EC GMP Guideline, Chapter 1, Quality Management

Let us take the examples of the regulations governing the interfaces between the individual departments, such as general prescribed conditions on SOP ("Mother SOP"), contents of instructions, performance of superdepartmental system audits, organisation of GMP training and importantly the implementation of laws and regulations into every department of the company.

Especially in smaller companies, it is possible for top management to deal with such regulations. However, top management often has more serious problems to deal with than the appearance and contents of standard operating procedures and/or the corresponding document numbering system.

In many cases, it is possible for interface problems to be clarified and perhaps solved by being worked through, as described in the GMP guideline (Chapter 2, Section 2.7), by the head of Production and/or the head of QC. In many cases, however, self-interest exceeds the common interest, preventing a solution which satisfies every side.

Let us look at another example: audits and in particular system audits. It is possible for these tasks to be performed internally, that is, by top management or by departmental staff. However, we know the usefulness of this type of work very well from experience. System audits run out of self-interest will be conducted only to solve the staff's own problems.

For many reasons it is difficult to imagine that a constructive audit can be done by the top management.

Nevertheless, it can be very helpful and beneficial to the top management to escort the system audits.

One possible solution is for the tasks to be performed by other departments or other personnel. However, even in this case it must be considered whether there will be any significant benefit from the work performed. After all, no one wishes to be scrutinised by colleagues with whom one has problems, and why should these colleagues be given yet another reason to publish these problems. As a result, an attempt is often made to present the department and its work in such a way that there appear to be no problems or that the cause of the problems lies elsewhere. Facts are concealed or given a "spin". Consequently, the audit is not performed with the necessary thoroughness and openness, and the conclusions derived from it are not worth the paper on which they are written.

Let us now consider the task of observing prescribed rules, laws and guidances.

At the moment, the pharmaceutical industry is being flooded with regulations, laws, directives, guidelines, commentaries and publications, and it is consequently very difficult to keep an overall view. It is not sufficient just to archive these documents carefully. Rather, because they represent the forefront of science and technology, they must be read, understood and implemented in a way beneficial to the company.

Implementation in the company often involves the inclusion of several different departments from different fields. In this case it is necessary to establish a coordination office that works out and performs the necessary implementation.

Two important examples of implementation of prescribed conditions that require company-wide involvement of several departments and services are "Change Management Systems" and "Validation of Processes".

We must now ask what alternative options exist, apart from intervention by the top management, to get around the problems described briefly in the examples given.

BUILDING UP OF NEW SPECIFIC SYSTEMS

Intradepartmental Structure with an Active Support by the Senior Management

With the aid of clear organisation and instructions, procedures, and prescribed conditions from the top management, it is possible to specify and document a department-specific system. According to these regulations, the QA System can be easily documented and proofed only for the department itself.

The advantages of such a system include the following:

- The individual system is clearly structured.

- System responsibility is spread over many shoulders and eventually peaks in top management.

- The documentation of the individual systems is easy.

- Maintenance is easy.

- Procedures may be aimed at individual departments and workers.

The drawbacks, however, predominate as far as a uniform company-wide system is concerned, as per the following examples:

- The individual systems are rigid.

- The system can act only reactively and not proactively.

- Nobody feels responsible for the entire system, as required by GMP.

- It can be established and thus function) only within departments.

- Interfaces are hardly regulated, if at all, and require instructions from "above".

- Solutions are and will always be intrinsically insular.

Team of Representatives Drawn from Different Departments

In order to combine the insular solutions and to apportion the responsibilities, the next step is to form a team of representatives from the individual departments.

Unfortunately, the advantages of this approach are not as great as could be imagined. These advantages include the following:

- The interfaces are more clearly worked.

- The system responsibility lies on many shoulders.

- The system care is more concentrated.

- In many cases the system can react preventatively.

- The company-wide QA concepts are more deeply rooted.

It must still be said, however, that the drawbacks only partially justify rational and effective establishment of this variation. These include the following:

- Preventative work is still difficult.

- Because the individual members of the team belong to different departments, there will be departmental restraints.

- The reward for making decisions is small because the responsibility of the individual members is delegated to the team.

- Decisions about insular solutions must be agreed upon.

- The authority to issue directives is very limited, if present at all.

- The question of who is "first among equals" arises.

Person or Department Responsible for the QA System

In order to get around these problems, as well as others not mentioned here, the DIN EN ISO 9001/4 provides some assistance and requires a management representative of the company to be responsible for the clear and proper functioning of the QA System.

In order to perform these tasks properly, DIN EN ISO requires that this person be a member of the management executive, and thus have a hierarchical position nearly equal to that of the department heads.

In this persons job description, the requirements and possibilities for a proper running of his or her daily work are clearly defined.

The job description should be known by the top management and the heads of the different departments.

There are fundamental advantages to this, a few of which are as follows:

- Uniform quality assurance concepts, and thus a company-wide view, can be established.

- Accordingly, a uniform and company-wide quality assurance system can be established in all the departments.

- The person responsible is not allied to any particular specialty or group, and is answerable only to the top management.

- In the event of problems, direct implementation of necessary measures can be endorsed by the active support of the top management.

- Problems can often be discussed in confidence, and can in many cases be solved before they become serious.

- It is possible for the responsible person to act as an arbitrator, in the event of problems and discussions between individual departments.

- Both internal and external data concerning the existing QA System can be gathered more easily.

The drawbacks to this approach are few, but on detailed discussion can become very explosive and lead rapidly to the failure of the person or department. These include the following:

- The person/department has a very high level of responsibility.

- In order to reap optimum benefits, there must be a high level of personal acceptance in the company.

- Specialist competence is important and must not be underestimated. The person/department responsible must be *au fait* with all the aspects under examination.

- In order to keep the system functioning properly, it must be clear that tough and isolating decisions must sometimes be made.

Biotest AG decided in the beginning of the 1990s to introduce a department dealing with the implementation and extension of a company-wide QA System based on individual GMP-related systems in the Production and QC departments. Similar structures are now in place for other companies (Schmidt 1997; Krüger 1998).

Before the final decisions are made about establishing a QA System, the reasons and goals of the system should be considered (see Table 7.6).

QA Systems can only be introduced by a company-wide decision. Based on the national and international regulations and laws, there is an obligation to employ a company-wide system, because many points within the regulations can only be introduced in a uniform manner.

This person or department should always be independent of manufacture, QC, sales and other departments. Also, according to DIN EN ISO 9001, this person(s) should be under the control of and answerable to only the top management. This means that a department or structure must be established in the company apart from all the other GMP-regulated departments.

The establishment of a QA Structure for individual departments or services is sufficiently ensured by the GMP requirement of the responsibilities of the head of Production and the head of QC.

The top management itself, whose active participation is required, could always, as described above, perform its duty sufficiently and still document its active participation by receiving continuous reports from the person in charge and by evaluation of these reports.

Company-wide QA by a person or department, nevertheless, does not mean that the existing responsibilities are weakened or rendered ineffective, according to AMG, PharmBetrV and GMP (Völler 1997). Instead, these activities and responsibilities become part of the overall functioning system and represent areas within it.

A few fundamental disadvantages associated with the selection and nomination of a person to take charge of QA should not be forgotten. In particular, the current or previous position held by the person within the company must be taken into account.

It would not be a good solution for a staff member from the Production or QC departments to be placed in charge of quality assurance, or even appointed representative of the top management for the entire company. Even if this member has already been active in the establishment of GMP, he or she will always be viewed as a member of the former department and therefore never achieve acceptance and support within the company. His or her partners will be, in most cases, the members of the lower and middle management levels.

Similarly, it is not advisable to nominate a colleague who will be leaving the company within a few years without building up a successor. As the establishment of a QA System is often an unloved task, the problem of implementation will be put off by personnel unless the conviction of its necessity has become sufficiently strong.

Table 7.6. Examples of areas for which a QA System can be introduced.

- For a department
  - production
  - quality control
  - research and development
  - maintenance
- For a certain product to reach a high quality standard
- For internal and/or external interfaces
- For services
  - company to costumer
  - department to department (interfaces)
- For the entire company to reach—based on regulations—a complete QA System
- For certification

It is thus always important to give the task to a person who has or will have the necessary acceptance in the company, is competent in many fields, and has the specialized competence to explain the necessary internal and external standards clearly and reliably in order to implement them to the benefit of the company.

TASK OF THE PERSON IN CHARGE OF QA

General Remarks

As can be seen from the definition of the infrastructure quality, the tasks of the person in charge of QA relate mainly to processes and coordination of interface problems. These include the three work areas listed in Figure 7.2.

The day-to-day work associated with each point requires many individual smaller tasks that in turn require implementation in individual departments such as Production, QC, Storage or even Research and Development (R&D) (see Annex 14 of the GMP guideline).

Let us examine the example of rules and guidance. A pharmaceutical company is affected by a large number of legal as well as "officially" voluntary standards. All of these points must be observed and implemented, some because they are law and others because they represent the state of the art in science and technology.

To do this, each individual department can work through these prescribed conditions and integrate them into their specific processes. In the ideal world, the individual departments or services would come together and try to find a common solution. In real life, however, the actual effect is usually that everybody attempts to shift a large part of the implementation to another department or impose their own interpretations.

One optimal solution to this problem would be to process the prescribed conditions by the independent office of the QA Department which can stipulate the individual tasks of the various departments and services.

The advantage of this method of processing is rapid and coordinated implementation with company-wide prescribed conditions, by QA in agreement with the top management.

Another example involves the performance of audits. Every now and then in a company, it happens that a batch cannot be released for a commercial product or processed to produce the desired end product. An audit is then necessary for two reasons: to establish the reason for the deviation from the specification, and to define measures to be taken as corrective and preventive actions.

The necessary problem audit can technically be performed by top management or other competent staff. However, it is difficult to imagine these persons possessing sufficient distance and objectivity to achieve the intended aim.

The auditing of individual services or departments by one another will also rarely achieve the desired solution to the problem.

Being independent of specialties and disciplines, the QA Department can find the reason for the deviation from the specification and work out a solution to the problem with the per-

Figure 7.2. Task areas of a person or department in charge of a QA System.

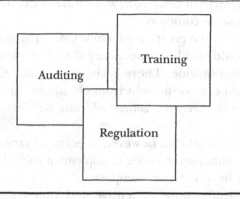

sonnel involved. It should be stated right from the beginning, however, that the aim of the audit must not be to apportion blame to individual staff members.

In addition to the general task fields of the QA Department specified in Table 7.7, the specific task fields are also mentioned.

Further company-specific points may be added, but these should be defined on a case-by-case basis by the company itself. In any case, it is important to draft a job description for the person in charge of QA. This job description should precisely describe all his tasks, authorities and competences, as well as the limitations to other responsibilities according to the national law.

First Steps for the QA Department in the Company

Following the implementation of the QA Department, the construction of a QA System is generally set up by the top management or the company board, with a desire to introduce it without great cost and a superficial structure that makes little or no substantial change to the company. It is often proposed that a known structure from a different company be implemented and introduced. The hope is that enough can be done to fulfill quality requirements without generating any major problems.

It can be said from experience that such ventures are doomed from the start, as it is well established that no two companies have the same infrastructure, procedures or personnel behaviour. Personnel whose activities are in external company branches know that different patterns of behaviour and thinking can also be seen even within the same company.

The expense of the creation of a new QA department within the company would also be unnecessary if such structures could be imported from outside. There is simply *no single* QA System that can be taken over or bought from elsewhere. Instead, *a QA System specific to the company concerned* must be developed and implemented.

It is definitely beneficial, however, to examine various systems from different companies in order to implement suitable elements and procedures for one's own company.

It is intrinsically more expensive and time-consuming to develop a brand new system, but this approach has two advantages. First it can be determined and shaped by the personnel themselves, and second, it can be tailored exactly to the conditions and requirements of the company.

However, as not every department or service in the company will be able to develop its own system, the QA Department in col-

---

Table 7.7. Specific task fields of the QA Department, which form the basis of the daily business of the staff of the department.

- Establishment and monitoring of a document structure.

- Cooperation of the establishment of quality policies.

- Coordination of validations.

- Coordination and performance of training.

- Coordination and performance of audits.

- Escort of the inspections by regulatory authorities.

- Assistance in the interpretation and implementation of requirements, guidances and laws.

- Cooperation in the project management, especially in research and development.

- Coordination of change management.

---

laboration with all the personnel of the company must specify the framework conditions for this system, so that there is a uniform understanding and a uniform basis for the QA System. The necessary department-specific and service-specific details can then be included as the system is developed.

To be able to impose framework conditions in the services and departments of the company, sometimes against the intentions of the personnel, one very important precondition should be met right from the start. According to Section 4.1.2.3 of the DIN EN ISO 9001, the QA Department and specifically its head should be a member of the management executive and, ideally, report directly to the company board or top management. Alternatively, it (or he) could be granted powers by the top management or company board. This is the only way to be sure that the authority to issue directives for the implementation of the proposed QA System is ensured, and that changes or reforms can be effectively introduced and purposefully established.

By stipulating the necessary competences and authorities in the corresponding job descriptions for each member of the department, the framework for the activities of the new department can be announced to the remaining management executive and effectively marked off.

Coupling a department to structures of the middle or lower management can result in insufficient thoroughness in the implementation of the issues that need to be introduced. In this event, the department is necessarily subject to departmental restraints.

BUILDING UP A QA SYSTEM

Analysis of the Current Situation

It is not only after building up a new structure for a QA Department that a company produces quality. For a long time, the company has prescribed conditions in the form of operating procedures, directives and laws in order to manufacture and produce products of high quality. It is an important aim of the QA Department, therefore, to add up all of these conditions toward the creation of an effective overall QA System.

The simplest solution here is to perform an analysis of the current situation. This form of analysis is called a *current analysis* or an *actual analysis* when it is compared with the desired or theoretical state. This method has the beneficial side effect that the personnel who may possibly be new) of the new QA Department become better acquainted with the structures and procedures of the entire company.

Selection of Regulations to be Applied

Before performing the analysis of the current situation, it is important for the QA Department to stipulate the intended overall structure of the QA System to be implemented. Two documents exist in Germany that contain the governing elements of QA. These are the PharmBetrV/GMP and DIN EN ISO 9001/4.

As a pharmaceutical QA System must follow primarily the pharmaceutical laws, rules and guidance, the question of the system's structure is easy to answer. It must correspond to the PharmBetrV and thus conform with GMP rules.

Other regulations may be necessary and obligatory in other countries (for example, CFR for U.S. companies).

What is being discussed here, if we follow only the PharmBetrV and GMP rules, is the establishment of *basic* requirements for a system that is sufficient according to currently valid jurisdiction (see Table 7.8).

As we shall see later on, consideration and initial implementation of some elements from DIN EN ISO 9001/4 is in many cases virtually indispensable to a company-wide coordinated procedure (including research and development, corrective and preventative measures, project management, specific points for the personnel and so on). Further elements from other regulations can then gradually be implemented as required. The mistake should not be made initially, however, of implementing all elements of the DIN EN ISO 9001 in addition to the EC-GMP guideline.

The stepwise combination of both QA Systems, which are in any case mutually supportive, then results in a *complete* or *comprehensive* system (Krüger 1998).

The pharmaceutical company has a duty to keep its level of knowledge and working practices up to date with the current progress in science and technology, as regulated in the AMG and PharmBetrV.

DIN EN ISO 9001/4 can without doubt be regarded as state-of-the-art technology. It has become a de facto worldwide standard and is also used very successfully in neighbouring fields (such as medical devices and *in vitro* diagnostics).

The successful implementation of the standard elements in a pharmaceutical company has already been described several times in the literature, and it is surely rational for further extension to follow this path (Pineda 1996; Krüger 1998).

It should, however, be stressed once again that certification according to the above standards is not implied and is not and has not been required (Auterhoff 1994).

Performance of the Analysis of the Current Situation Basic Requirements According to GMP/PharmBetrV

In order to satisfy the basic requirements of a QA System according to PharmBetrV and the EC GMP guideline, the generally valid catalogue of queries presented in Table 7.9, containing the most important points from the regulations in the form of memoranda, should be prepared.

This ensures that the most important points during the audit will not be forgotten, and that identical questions regarding procedures will be asked in all departments. More detailed questions can also be asked (Steinborn 1992). This method ensures comparability of the answers from the individual services and departments.

It should nevertheless come as no surprise that more data are available and documented in the individual departments than is presumed.

The departments of Production, QC, Product Storage and Maintenance should be selected as primary target groups for inclusion in the analysis of the current situation, as these are regulated in detail by the PharmBetrV and also the GMP rules.

Table 7.8. General summary of the basic requirements of a QA System, according to Chapter 1 of the GMP Guide and §1a of PharmBetrV, the existence of which must be proven by adequate documentation.

GMP/PharmBetrV
Evidence: Description and documentation of procedures in

- Production and QC

- Uniform documents and documentation structure

- Organigram and job description

- Delimitation of responsibilities

- Clearly defined contract with subcontractors

- Validated procedures and methods

- System of self-inspection for checking the observance of legal requirements

- System for processing of complaints and product recall

- Ensuring peripheral conditions

  - Personnel

  - Hygiene in all rooms and areas

  - Instruments and equipment

  - Starting materials

Comparison of Theoretical Conditions with Analysis of the Current Situation (Theoretical vs. Actual Analysis)

Before an analysis of the current situation is performed, it is essential to set a uniform understanding of the desired level of implementation of the regulations to be applied in the company. This should either be discussed with the department or given by the QA Department itself in advance, in the way of an SOP.

In this way, tacit agreement is generated right from the start on the desired extent of the company's own QA System. As an example of this, the document management should be described in the form of instructions or SOPs (Schmetz 1997).

It will surely have been established by the analysis of the current situation that many services or departments possess and circulate their own types of standard operating procedures. When deciding to reform this document management, the following questions should be borne in mind:

- Should the procedures have a uniform appearance throughout the company?

- Should a uniform numbering system be introduced?

- Should the procedures be administered and recorded centrally?

- Should the font sizes and types be stipulated?

If common answers dedicating a common understanding of the requirements were received, they can be concentrated into rules dealing with managing SOPs and/or documents.

During the run-up phase (comparison of the theoretical vs. actual conditions), it can be shown directly to the staff how to handle the regulated documents in the future.

Furthermore, in order to avoid lengthy and exhausting discussions during the run-up phase, they should also specify necessary points in the form of generally valid instructions, include the following:

- extent/depth of documentation;

- validation of methods and procedures;

- hygiene requirements;

- requirements for qualification of the personnel;

- self-inspections/audits; and

- training.

It is known that, in order to keep their options open, the individual members of the various services and departments tend to write and record as little as possible in the form of instructions.

Table 7.9. Basic questionnaire for the analysis of the current situation. The questionnaire is used in the beginning of the implementation of a QA System, according to basic requirements from PharmBetrV and EC-GMP.

- Is there an organigram of the department/service available?

- Is there a job description available for at least the responsible expert) personnel?

- Are the responsibilities, according to §19 and §63a of the AMG, clearly stipulated in the job descriptions or in any other documents?

- Is the release of the products clearly regulated?

- Is there a list of the available and applied standard operating procedures?

- Is there a list of quality-relevant equipment?

- Do these lists cover all points necessary for

  - manufacture

  - quality control

  - packaging and storage

  - hygiene (personal, room, instruments)

  - cleaning and qualification?

- Are valid specifications available for all the tests, such as

  - incoming goods

  - in-process controls

  - final testing

  - room hygiene

  - release (Installation Qualification (IQ), Operational Qualification (OQ), Performance Qualification (PQ)) of equipment?

- Are there written procedures observed, and are the processes properly documented?

- Have the applied procedures and methods been validated?

- Are the personnel qualified for their activities?

- Is there a procedure for recall of a product already on the market?

- Is there a training program?

- Is there a self-inspection system or audit system?

Detailed stipulations in the individual services and departments are then possible on the basis of the general prescribed conditions of the QA Department.

In addition to the findings of the theoretical vs. actual analysis, deficiencies must also be evaluated with regard to severity (according to law and in-house). It must thus also be stipulated from the start how individual deviations are to be evaluated.

Deviations from the theoretical are evaluated according to the prescribed conditions and entered into a measures catalogue for the individual department or service, and this catalogue serves as a basis for further work.

IMPLEMENTATION OF A QA SYSTEM

Implementation According to the Actual Analysis

On the basis of the theoretical vs. actual comparison and the evaluation of the individual deviations, a program of tasks should be drawn up that must be continuously met by the individual departments and services.

It is very important that the QA Department have the full backing of the top management in the stipulation of the tasks for the different departments. Otherwise, the implementation of a QA System will fail at this early stage.

In many cases it is advisable to structure the measures into urgent tasks (involving breaches of legal regulations) and less urgent tasks (involving inadequate implementation of legal and other regulatory stipulations).

Important assistance is thus given in the individual services and departments. Where this is not the case and there is only voluntary observance, the deficiencies list is worked through in order of increasing associated workload.

The QA Department must always be a responsive and *consistent* partner and give constant assistance, especially when new problems occur. To achieve this, it is necessary that the newly arriving legal and regulatory stipulations are continuously interpreted by the relevant service and specifically documented with proposals for implementation. Depending on their function, they can thus go beyond the coordination of tasks across services and departments required by necessity and prescribed conditions (interface problems), thereby preventing poor work repetition of tasks.

The QA Department must always be prevented from performing by itself any of the tasks on the deficiency list for which it has not been designated specifically responsible. The departments and services will very rapidly request that the QA Department issue worked-out proposals to which the personnel can turn. From here, it is only a small step to the work being actually performed by the QA Department itself.

Even when everything in a company appears to be running smoothly, it should never be assumed that no work is necessary.

During the work on the deviations listed in the audit report, the goal must be defined very clearly and never changed unless truly necessary. It is not helpful to add more and more points unless they provide a deeper or better solution.

If, for example, the goal is to add a missing SOP, it is not worthwhile to ask the person constructing the SOP to add additional SOPs. This question will frustrate the person. It is best to complete and implement the missing SOP first, then address in the audit report the necessity for additional SOPs.

For the motivation of the staff who must fill all the measures mentioned in the audit report, it may be helpful to give an occasional statement that describes the ongoing work to the head of the department or to the top management.

If two departments are dealing with similar issues or contemplating similar measures, they should have a meeting on this topic under the leadership of the QA Department.

It should also be beneficial to inform the person in charge that he or she will have access to information or help from other departments inside the company which have already reached a solution.

But two important points must be taken into account: first, the person seeking the information should not be treated as a pupil who has a lot to learn; second, the department should give willingly the information.

Motivation of the staff (see also Section 11.2.2) to work on the measures based on the report is the most important point beside the full backing of the top management.

If the personnel is absolutely unwilling to work on the deviation, then the QA Department can try two approaches: first, asking the top management for a helping hand; and second, working on motivation. Motivation means that the QA Department should explain why the point mentioned as deviation is important and needs to be worked on. The reason may be given by law, in which case there will be no discussion on it. The reason can also be given by exploring what negative effects will arise for the company or the quality of the product if this work is ignored. This is the best way to motivate the staff because they will really understand the problem and the solution.

The very last step will be to ask the top management for a helping hand!

Some Differences Between EC and U.S. Points of View on QA

In the United States as well as in the EC, basic regulation on in the manufacturing of pharmaceutical products is provided be the GMP regulations issued in 1963 by the FDA and revised in 1969 by the WHO.

The authorities have transposed these regulations into national law as the Code of Federal Regulation (CFR) controlled by the FDA, and the European Directive 91/356/EEC controlled by the EC in a somewhat different context.

The content of the GMP regulations are nearly the same in both jurisdictions, and differ only on certain issues.

In each European country, the directive has to be adopted into national law by organisations such as the PharmBetrV in Germany, together with the GMP guideline III/2244/87.

The QA System set forth in the WHO GMP guidance as well as the European guidance is described as a concept or an arrangement implemented to ensure that pharmaceutical products are of the quality required for their intended use. It should be well documented and all departments dealing with this concept should be staffed with competent and sufficient personnel and be equipped with suitable equipment and premises.

In 1996, the FDA issued a document dealing with the interpretation of CFR 210 and 211 (Amendment to 210/211 CFR 1996) and the understanding of the corresponding GMP regulation still in draft). The special part 211.22 describes the position of the QA Department, which is understood in its function to be the same as the German QC Department. When the FDA speaks about the QA Department, it always means, in the sense of the German law, the QC Department.

The difference—based on the U.S. law—between *QC* and *QA* is recognised to be operational. The QC unit is part of the QA unit, and has the functional duty of performing the tests and assuring that proper specifications and limits are adhered to.

On the other hand, many German and other European pharmaceutical companies have started to concentrate the responsibilities for the proper function of the quality system based on Chapter 1 of the EC GMP regulation within a QA Department. This department has *no* responsibility for the *quality of the products* and the quality of the environmental conditions.

The responsibility for the quality of the products is delegated to the head of Production and the head of QC.

One more important point has to be mentioned.

According to the European regulation 75/319/EEC article 22.2, *one* qualified person has to guarantee that every batch released has been manufactured and controlled according to this regulation and met all specifications before coming into the market. This regulation is in agreement with the FDA's cGMP regulation. Both demand that the release of a pharmaceutical product be done by a responsible member of the QA Department.

The basic qualifications for this person are mentioned also in Chapter 23 of the regulation.

The German authorities have not transposed this part completely into national law. The PharmBetrV declares in §7 (Release) that a pharmaceutical product can be labeled as a released product only if the documentation of the quality control and production have been signed.

Previous to giving a release signature, the head of QC, together with the head of Production, must evaluate the batch records (EC-GMP Chapter 2.6, II), and the monitored and controlled environmental conditions (EC-GMP Chapter 2.7 and Chapter 6.3).

The meaning of a QA unit is therefore somewhat different in its understanding and function from a QC unit, as laid down in the CFR regulations and amendments (Wenzel 1995; Immel 1998).

EXTENSION OF THE QA SYSTEM

Further Development

Monitoring and support by the QA Department during the work on the deficiencies listed can be used to extend the QA System step by step on an ongoing basis.

One necessary item is to perform further training of relevant staff during the implementation. Further audits will be necessary to perform more thorough working on the system once it is implemented. Even partial success should be documented and reported to the management, as this can be a psychologically valuable aid. Only in this way can the personnel be motivated to extend the system and dedicate themselves to additional proposed tasks.

The increasing interplay between audits, training, interpretation of rules and guidance and stipulation of still missing QA structures gradually creates a system that is maintained by the QA Department and extended with the cooperation of relevant departments and services. To do this, the QA Department should issue instructions and standard procedures that repeatedly specify audits, self-inspections, or training sessions at fixed time intervals, in order to maintain and further improve the current quality of work.

In this context, it is also important that particular consideration of the implementation be given to the interface problems. Again and again, emphasis on management between the departments and services involved is indispensable for the extension of the QA System. It is well proven that most problems in a company originate from poor attention to the cooperation between different departments and services.

Here, it is the task of the QA Department to construct, for example, a client-supplier relationship inside the company that takes account of the basics of DIN EN ISO 9001/4. Element 4.3

(contract review), in conjunction with element 4.6 (purchasing) from the above mentioned standard, gives clear instructions as to how a system should be formed. For example, if Production wishes to have analyses of intermediate products performed by QC, particular attention should be paid to ensure that all necessary information is passed from Production to QC and vice versa, so that the operation runs smoothly without further extensive requests for information.

Furthermore, a system must be implemented that gives the information to the QC Department in advance when a sample has to be proven in case of time management. Especially for in-process samples, it is very important that the results be given back as quickly as possible, because of the lag time during manufacture.

A further example is the transfer of new products from R&D into the Technical or Production Department. This is a very complex procedure, but it can be governed by the two standard elements described above, 4.3 and 4.6, according to DIN EN ISO 9001 with respect to 4.4 of the standard (see Section 11.1.2).

Unnecessary question-and-answer games in the company are not economical and must be strictly avoided in all situations. The creation of a internal client-supplier relationship according to DIN EN ISO) at interfaces establishes a level of communication that generates clear information flow in both directions.

In general, it can be stated that the resulting necessary interface procedures establish what information exchange and specific responsibilities are required for economical processing and working.

It is thus absolutely essential that, following the initial implementation of a QA System in individual services and departments, it is checked for compatibility with other systems in the company and adapted if necessary.

This can be done only by an independent office, a QA Department or by audits. One type of audit that has proven useful here is a retrospective product audit starting at the sale/storage of a finished medicinal product and going backwards to the first use of a starting material or, even better, to the purchase of the starting material. Only in this way can the interplay of the individual systems and in particular the management of interface problems be best tested.

Particular attention should be paid here to the timeframes for individual tasks and procedures, and to the drawn-up list of failings to be corrected.

Another important point is that the proper functioning of batch recall should be specified in collaboration with the person in change of the complaints, and should be audited in detail.

In summary, it can be said that the extension of a QA System is always performed cyclically, as shown in Table 7.10.

IMPLEMENTATION OF SOME RELEVANT ITEMS OF THE DIN EN ISO STANDARDS

Extension Based on DIN EN ISO 9001

QA Systems that are constructed according to DIN EN ISO 9001 with the help of DIN ISO 9004 contain all 20 elements of this standard. By comparison, the EC-GMP rules contain only 9 chapters.

Nevertheless, as the pharmaceutical industry possesses more thorough regulations (Auterhoff 1991) in all elements, it does make sense in the final analysis to incorporate DIN EN ISO when elements give helpful guidance, and when more stringent regulations cannot be found in other prescribed conditions such as national laws and rules.

Above all, they must support the available and stipulated rules, guidances, and structures, and be clearly able to be introduced into a pharmaceutical QA System.

As described previously, the QA Department continuously aids the individual departments and services in the extension and further development of the QA System on site.

Where needed, the QA Department can provide additional and helpful supplements from DIN EN ISO 9001. However, as with the implementation of points from the PharmBetrV and GMP, it should be agreed from the start which elements should be implemented and to what extent, because not every element can be applied equally to all services and departments.

The possible presence of some already existing regulations can, as in the previous case, best be established in a theoretical vs. actual analysis, such as an audit (Steinborn 1992).

From practical experience, the implementation of the elements listed in Table 7.11 has proven to be quite useful. They ensure practical front-line support of the basic QA System.

It is surely advisable for the system to include other departments or procedures such as Marketing and Sales, Clinical Research, Registration and even the Technical Department. Owing to the complexity of the necessary tasks, this should not be a subject for the QA Department which is still in the early stages of its development.

Responsibility of the Management

More thorough implementation of this element (which here comprises the responsibility of the management *and* the management review) should have the effect that the top management level is more actively involved in the awareness and support of the system.

It makes no sense to try to prove the existence of a functioning QA System by pointing to the QA Department, or to pass the buck by saying "it's not my pigeon". Top management must become very actively involved in the process in the form of data and stipulations such as awareness of the main contents of the audit records and the complaint statistics, as well as the active stipulation of quality targets and monitoring and support of corrective and preventative actions.

It can be very helpful for the performance and implementation of necessary measures if the head of the QA department is represented on the governing committees or boards of the top management, and is every now and then asked to make a brief statement on the progress of the work.

It can also be of benefit for both parties if the management is given training on GMP or relevant rules and guidance, and the staff is trained to understand the implementation of the QA System. The training for the staff should ideally originate from the management level.

Table 7.10. Basic schematic procedure for the implementation and extension of a QA System as a cyclical procedure.

1. Determination of rules and guidance to be applied

   • Checking/adaptation of existing systems

2. Structuring of the system to be implemented

   • Determination of the services, departments and procedures to be included

   • Determination of the extent of the system and the depth of implementation of the rules to be applied

3. Drafting of a questionnaire

   • Determination of the queries and themes to be answered

   • Determination of the evaluation criteria for correspondence and deviations

4. Performance of the analysis of the current situation

   • Determination of the deviations

   • Drafting of a working catalogue

5. Implementation of the necessary measures

   • Elimination of the deviations found

   • Implementation of the required measures

6. Implementation/development of the system

   • Test for internal compatibility

   • On-going checking audits) and training

7. Return to Point 1.

Only in this way is it ensured that the meaning and especially the benefits (which are without doubt present) of a QA Department are made transparent to the company.

Design

Due to the increasingly cost-intensive development of new pharmaceutical products, the observance of strict schedules up to, for example, registration and marketing is an essential assumption.

Annex 14 of the GMP guideline contains a comprehensive chapter dealing with the manufacture of samples for clinical trial.

But prior to the preparation of clinical trial samples, the Research and Development Department has performed a lot of work over a long period, with the aim of creating a licensing dossier or other documentation relevant for approval.

Table 7.11. Extended, primarily necessary requirements of the QA System, according to DIN EN ISO 9001/4 construction of a comprehensive QA System on the basis of the already existing structures according to PharmBetrV and EC GMP.

Elements according to DIN EN ISO 9001

• Stipulation of the tasks for the responsibility of the top management

• Evaluation of the Quality Assurance System by the top management

• Introduction of project management/design steering

• Extended system of complaints and product recall, correction and preventative measures and change management

• Systematic determination of the training requirement

• Statistical methods

Elements according to DIN EN ISO 9004

• Economic consideration of all procedures to be established.

• Staff motivation

   • Transfer of responsibility and competence

   • Inclusion in the information flow

   • Training

• Product safety/product liability (optional)

The establishment of a project management that accompanies the new medicinal product to market release makes it possible for a documented course to be stipulated (and include the elements of the design according to Element 4.6 of the DIN EN ISO 9001), and be tracked by milestones (of which the manufacture of samples for clinical trial represents just one milestone).

It is most important to say that an early GMP-conforming production of the product for clinical trial must be ensured, and that this cannot be done only by observance of the DIN EN ISO.

Paying attention to both regulations will finally result in the product's acceptance for clinical trial.

It should also be mentioned that the DIN EN ISO standards pay in many points particular attention to the problems of interface regulation, EC GMP regulation, and particularly Annex 14 regarding manufacture and QC.

A very good procedure (client-supplier relationship) governing the transfer of results from one service/department to another service/department (Research → Technical laboratory → Production) can be worked out from the elements design output, design review, and design verification.

Furthermore, it is possible for all the documents necessary for registration to be drafted at an early stage.

Corrective and Preventive Actions/Change Management

During the actual analysis, deviations from once established procedures or from the necessary prescribed conditions will surely have been noted.

The individual problem items should be worked out and listed on an action list, and thus hopefully eliminated.

The aim of the implementation of this point should be to establish a system that *proactively* ensures that the eliminated deficiencies are extremely unlikely to recur in the future. When a mistake is recognised or occurs, *active* efforts should be made to prevent recurrence of the mistake. The mistake and its possible cause must be examined closely and solutions found and implemented.

It is not sufficient just to be satisfied that the mistake has been corrected and the responsible member of staff has been educated.

In this context, it is of benefit for the QA Department to participate in the relevant department or service to discuss individual points from audits or meetings.

In some companies, Quality Circles that deal only with the working through and elimination of problems have been established. These discussion circles can be chaired by the QA Department.

The items in a corrective and preventive system should include the following:

- increased frequency of training;

- drafting of an operating procedure;

- reorganisation of procedures; and

- expansion of communication or documentation.

Training

On the basis of the motto "Only an informed member of staff is a competent member", a training program should be written every year and should stipulate how and in what subjects the staff should receive further training.

In accordance with a written operating procedure drafted by the QA Department, the responsible manager, in agreement with top management, should together with the personnel draft a training plan at the beginning of the year. This plan should clearly lay out in what subjects further development of the personnel is required and should be aimed for.

Training should then take place not only internally. External training has multiple psychological advantages, not only because it gets around production facility blindness, but also because it can be used to demonstrate the value of personnel (see also Element 4.2.1 of the standard DIN ISO 9004).

The QA Department then has the task, possibly together with the Training Committee, of monitoring the desired training (Schrätzenstaller-Rauch 1997; Stepfan 1995; Dietrich 1997; Immel 1998).

Note: The motto must also be applied to the management levels.

Statistical Methods

Statistical recording of data and here particularly of production data is demanded by the GMP guideline (Chapter 5, 5.8).

Further quality-relevant data should also be compiled and systematically evaluated throughout the year.

The points to be evaluated here include the following:

- yield calculations;

- trends in analysis;

- trends in the manufacturing process;

- change control;

- number of and reasons for unsatisfactory batches;

- environmental monitoring;

- statements from the graduated plan process;

- internal and external complaints; and

- failure and repair of equipment.

A corresponding operating procedure based on DIN EN ISO 9001, Element 4.20 should stipulate which data sets should undergo statistical analysis. The statistical evaluation of corresponding quality-relevant data should be dealt with.

This is a very good way of recognising trends that, by corrective and preventive measures, allow the inclusion of management levels to be evaluated and worked through.

Extension on the Basis of DIN ISO 9004

Finally, two important points should be mentioned that should be discussed at an advanced stage of the establishment of a QA System.

Economic Considerations

The performance of all tasks and the establishment of changes or reform should always be considered and discussed against the background of economic and business considerations for the company.

Everything may seem to be very well regulated (particularly complex procedures and regulations across interfaces) so that the likelihood of a problem occurring is reduced to practically nil. As a consequence, however, the personnel may no longer have any freedom in their work.

In many cases, for example, the 75 percent solution with a justifiable residual risk can be introduced far more reliably and economically than by forcing staff toward 100 percent safety.

If rules are too restricting, the motivation of the personnel and their individual sense of responsibility is very rapidly destroyed, and they become "Jobsworths" working strictly according to instructions. The QA Department is then made the scapegoat for any errors that occur: "It's their job to tell me what to do".

It is important to first create a rule and then, in the course of time or periodically with the aid of a quality audit, check whether it can if required slowly be adapted to the most practicable solution. The 75 percent rule also states: better incomplete and right than wrong in detail.

The economic considerations of a measure should always be given high priority in further tasks and implementations.

Together with the 75 percent rule, running audits and the following reporting of the deviations observed should be discussed. The demand of new or changed requirements should take into account the economical balance between legal recommendations. Each new or changed regulation mentioned in the audit report must give sufficient—or if there are real problems on the quality of the product—more safety on the quality of the product.

One should prevent overregulating the work.

For example, during the audit is has been observed that the maintenance of the autoclave is done twice a year in order to check the temperature and the pressure.

The auditor may propose—based on his or her own feeling—that the equipment should be controlled four times a year. Because there is no regulation dealing with this problem, the economics and advantages of the proposal must be discussed.

Motivation of the Personnel

In principle, no particular attempts need to be made to motivate the personnel in a company that has implemented and maintained a QA System according to the regulations, as their cooperation is in almost all cases actively promoted. They are thus able to form the system themselves under the leadership of the QA Department (Wilhelm 1997; Crostack 1997).

In contrast to what is stated in DIN ISO 9004, which proposes motivation exclusively in the direction of enhancing quality awareness and creation of quality yardsticks, a few other important factors should be mentioned (see Table 7.12).

The drafting of job descriptions with accurately defined limitations in the areas of responsibility and competence creates a clear basis that stipulates the framework for specific activities. It provides sureness during work and describes a specific working field in which the personnel feels free to make decisions.

The ability to make decisions and to take responsibility for these decisions within the framework of activities is an important motivator in the daily work. In addition, the receipt of further information outside the working field rounds out the ideal of an informed and thus competent staff. The staff is motivated because they feel their superiors have confidence in their work and they can often follow their own initiatives.

One example of the transfer of responsibility is the drafting of operating procedures. If a staff member writes these SOPs himself and makes them valid with his own signature, these documents are *his* procedures. He alone is responsible for the content, accuracy, and practicability of what is written, and thus responsible for the SOP.

Another boost to motivation can be achieved if, during the course of the determination of the training requirements, staff members have input into the direction of further training and extension of experience. In collaboration with the manager, a training program should be drawn up and followed through.

When performing in-house training, it may sometimes be of benefit for staff who have attended an external training course to report on this in an internal seminar and thus pass on knowledge to their working colleagues.

STRUCTURE AND RESPONSIBILITY OF THE QA DEPARTMENT AT BIOTEST AG

Position of the Department

Beside all the other departments which are necessary in manufacturing pharmaceutical products, Biotest AG began in 1989–90 to build up a new department called *Zentrale Qualitätssicherung* (Central Department for Quality Assurance Affairs).

At this time it was a new step for a pharmaceutical company to create such a department. Before that time, it was up to the Head of Production and/or QC to deal with the implementation and observance of national and European laws.

But due to the fact that parts of Biotest AG deal with blood components, it was very helpful to have one department guide all the other departments in running their work on virus validation and discuss process validation as well as validation of analytical methods.

It also appeared to be necessary to introduce a global (company-wide) QA Structure.

As mentioned earlier, Biotest AG has had inside its different departments a basic QA System according to GMP and PharmBetrV.

Therefore, one major goal was to join these basic systems together into a global QA System valid for the entire company.

Table 7.12. Suggestions for the motivation of the staff to support the QA System for the entire company.

- Drafting of a job description, together with the transfer of responsibility and competence.

- Forwarding of all information—not just task-specific.

- Active assessment of the training needed.

- Responsibility for the drafting of documents.

- Active inclusion in solutions of quality problems.

- Acceptance as equivalent working and discussion partners at every level.

As we know from our own work, it is not easy to withdraw the responsibility for the QA System from the head of Production and the head of QC and place it in separate hands. Furthermore, this new department will have the power to give orders to these personnel and their departments in matters of QA.

To acclimate all of the company personnel to this new situation while giving the QA Department the support necessary to do its job properly, the new department was directly linked to the board of directors.

Every member of the new department has a job description wherein, beside other important points, the following will be mentioned:

- They may read every document (exceptions are mentioned).

- They may ask questions of/speak to all personnel.

- They may enter every room in the company.

These three points are very important for the daily job of acquiring information and discussing problems even without the assistance of the head of the department.

The head of the QA Department also has in his job description the duty to inform the board of directors about problems he can't solve on his own, or which are important enough to be addressed at the meeting of the board of directors.

All the job decriptions of the members of the department are signed by a member of the board of directors.

The QA System is a growing system and everybody in the company has a duty to help the QA Department support and expand this system. For this reason it is necessary to have a periodic review of the system to inform the board of directors. At Biotest AG, this is done either by a written report or during the meeting with the heads of the departments, not less than biweekly.

General Responsibility

Biotest AG has different pharmaceutical fields, including:

- *in vitro* diagnostics such as testing systems;

- pharmaceutical products such as intravenous applications; and

- medical devices (such as those for medicinal use).

The QA Department of Biotest AG consists of three personnel. Every member has his own field to work on for implementing his specific regulation. That means each pharmaceutical field will have its own QA system that will need to be implemented according to the specific laws.

Rules that manage the overall system for Biotest AG will therefore be discussed and finalised between the member of the department. Based in this work it can be assured that in new regulation, such as SOPs, no violation of any special legal regulations will occur.

It was also stipulated in the job description that each member of the department has the ability to convene meetings and to request that the staff or the heads of the different departments attend.

Included in the job description of the head of the QA Department is the ability to introduce regulations directly into the partial or overall system in order to enhance the QA System. Theoretically, this can be done without the assent of the heads of departments or the board of directors. To avoid prolonged arguments, however, this should never be done.

Following is a short example of how to introduce a regulation such as a general SOP when different fields are involved.

Introducing a General SOP

A member of the QA Department who plans to manifest a new SOP must enlist at least one member of each field—or asks the head of the field to select a member—to give a helping hand on discussing and writing a first draft of this SOP. At the first meeting, a very rough draft is given by the QA Department for discussion.

After the members of the meeting finish the first draft, it will be sent to department heads to discuss within their fields and provide feedback within a specified time. The deadline should be stressed and no late responses should be accepted.

The members of the first meeting should reconvene to discuss the departmental feedback. The final draft will then be sent to the department heads and the board of directors for their final comments on the SOP.

If necessary, an explanation should also be given to the heads of the fields as to why all of their comments have not been added. For the board of directors, a note should be added regarding why this SOP should be implemented and what advantages will arise for the company.

After finishing the SOP, the QA System of Biotest AG requires that an author (who is responsible for the content of the SOP), the head of every field involved or one member of the board of directors, and finally the QA Department sign the final valid SOP.

These signatures, together with the responsibility specified in the SOP, give management responsibility for the QA System at Biotest AG.

Detailed Responsibility

Besides the general responsibility of the QA Department to enhance and support the QA System of Biotest AG, each member of this department has his own detailed responsibility in the different fields he works for.

The main duty of the staff is to introduce and translate the necessary regulations for each field into the daily work of the personnel of the field. This includes the following tasks, where applicable:

- conducting audits for continual analysis of the current situation;

- training personnel;

- introducing general and specific new regulations, such as the European regulation for *in vitro* diagnostics (CE marking);

- ensuring certification according to DIN EN ISO 9001;

- dealing with virus safety and process validation;

- registration of *in vitro* diagnostics with national authorities such as the Paul Ehrlich Institut (PEI); and

- signing all SOPs for the implemented QA System.

One final point should not be forgotten. If you have a very clever secretary as we do, most of the paperwork such as archiving of laws, directives, guidelines, votes and so on can be done by the secretary in a computer-assisted system. Thus, it is easy for everyone in the QA Department to get quick information on the latest standards and regulations.

REFERENCES

AMG. 1996. Gesetz über den Verkehr mit Arzneimitteln, pharmind Serie Dokumentation. [Arzneimittelgesetz (German Medicines Act), Law on trade in medicinal products, pharmaceutical industry series documentation.] Editio Cantor Verlag.

Amendment to 21 CFR 210/211. 1996. Amendment of certain requirements for finished pharmaceuticals. Proposed rule.

Auterhoff, G., 1991. Qualitätssicherung in der pharmazeutischen Industrie *Pharm. Ind.* 53 (9)1–3.

Auterhoff, G. 1994. Qualitätssicherung in der pharmazeutischen Industrie *Pharm. Ind.* 56 (9):785.

Crostack, H.-A., J. Floel, M. Freitag, and M. Maas. 1997. Maß für Motivation. *QZ* 42:168–172.

Crostack, H.-A., J. Floel, T. Pfeifer, S. Korsmeier, and P. Flak. 1997. Mittel zur Motivation. *QZ* 42:1361–1364.

Dietrich, R. 1997. Entwicklung und Realisierung eines Computerbasierten Lernprogramms zur GMP-Schulung. *Pharm. Ind.* 59(11)941–945.

DIN EN ISO 9001. 1994. Modell zur Qualitätssicherung/QM-Darlegung in Design, Entwicklung, Produktion, Montage und Wartung, [Model for Quality Assurance/Quality Management considerations in design, development, production, assembly, and servicing] Berlin: Beuth-Verlag GmbH.

DIN EN ISO 9004-01. 1994. Qualitätsmanagement und Elemente eines Qualitätssicherungssystems, Leitfaden. [Quality Management and elements of a quality assurance system, Guide] Berlin: Beuth-Verlag GmbH.

EG-GMP. 1995. Leitfaden einer guten Herstellungspraxis für Arzneimittel. [Guide to good manufacturing practice for medicinal products.] 4th ed. Aulendorf: Editio Cantor Verlag.

EG-GMP. 1995. Richtlinie in: EG-GMP Leitfaden einer guten Herstellungspraxis für Arzneimittel. [Directive in: Guide to good manufacturing practice for medicinal products.] 4th ed. Aulendorf: Editio Cantor Verlag.

EWG Richtlinie Band I, Juli 1994. 75/319/EWG, Die Regelung Arzneimittel in der Europäischen Union, Band I, Juli 1994. [EEC Directive 75/319 Rules governing medicinal products in the European Community, Volume I, July 1994].

EWG Richtlinie 91/356/EWG. Die Regelung Arzneimittel in der Europäischen Union [= EEC Directive 91/356/EEC; Rules governing medicinal products in the European Community.]

Häusler, H. 1993. Quo Vadis—Qualitätssicherung. *Pharm. Ind.* 55 (10):884–889.

Immel, B. 1998. Meeting requirements of the European Unit. *BioPharm* (January), 46–49.

———1998. The essence of training Part 1, *BioPharm* (May), 90–100.

———1998. The essence of training Part 2, *BioPharm* (June), 62–70.

Krüger, D., I. Fohmann, A. Hardtke, U. Siegert, and A. D. Little. 1998. Internationales Qualitätsmangement in einem pharmazeutischen Unternehmen. *Pharm. Ind.* 60 (2)105–108.

Mühlenz, E. 1992. Qualitätssicherungssystem des EG-Leitfadens— mehr als Qualitätskontrolle? *Pharm. Ind.* 54 (9):773–776.

PharmBetrV. In 1995. *EG-GMP Leitfaden einer guten Herstellungspraxis für Arzneimittel.* [Pharmaceutical Concern Ordinance in *EC-GMP Guide to good manufacturing practice for medicinal products*, 4th ed. Aulendorf: Editio Cantor Verlag.

PIC 1/97. 1997. Guide to good manufacturing practice for medicinal products.

Pineda, S., E. Dietze, G. Brendelberger, and O. Schmidt. 1996. Einführung eines unternehmensweiten pharmazeutischen Qualitätssicherungs systems. *Pharm. Ind.* 58 (10)889–896.

Schmetz, A. and A. Maas. 1997. SOP Standardarbeitsanweisungen. *Pharm. Ind.* 59 (11)934–941.

Schmidt, O, 1997. *Pharmaceutical Technologie Europe* (Nov.) 9 (10)40–44.

Schrätzenstaller-Rauch, B., A. Maas, and A. Schmetz. 1997. GMP-Ausbildung im Betrieb. *Pharm. Ind.* 59 (6)476–480.

Steinborn, L. 1992. In: *GMP ISO 9000 Manual, Quality Audit Manual for Healthcare Manufacturers and their Suppliers.* Interpharm Press. ISBN 0-935184-34-1.

Stephan, P. 1995. *QZ* 40:380–384.

Völler, R. H. 1997. Definition von Qualitätskontrolle, management und sicherung. *Pharm. Ind.* 59 (8):695–696.

Wenzel, J. 1995. Arzneimittelproduktion in den USA. *Pharm. Ind.* 57 (2):121–129.

WHO GMP Oeser/Sander Pharmaceutical Concern Ordinance. Stuttgart: Wissenschaftliche Verlagsgesellschaft mbH. 1997.

Wilhelm, H., 1997. Engagierte Mitarbeiter sind Gold wert. *QZ* 42:948–950.

8

Implementation of a Quality System: A Report of Practical Experience

Oliver Schmidt
Concept Heidelberg
Heidelberg, Germany

This chapter describes how the requirements of Good Manufacturing Practice (GMP) and International Standardization Organization (ISO) 9000 can be concretely established in a pharmaceutical Quality System. It also explains how, after the establishment of the Quality System in accordance with ISO 9000 and European Commission (EC) GMP, the system can be expanded to include two additional requirements. The two expansion steps are:

- Food and Drug Administration (FDA) compliance; and

- Integration of environmental management requirements in accordance with ISO 14000 and the EC Regulation 1836/93 (Eco Audit). Both these steps have been successfully completed.

PROJECT DESCRIPTION

The Enterprise

Schering Produktionsgesellschaft arose from GEHE Medica Produktionsgesellschaft which, in turn, was founded out of Jenapharm in mid-1995. As a contract manufacturer, Schering GmbH & Co. Productions KG currently produces solid dosage forms at its Weimar site and sterile medicinal products at its Jena site (Germany).

The company now has 320 staff members at its Weimar site and 120 staff members at its Jena site.

The Requirements of the Management

The management stipulated that the Quality System to be established should comply with the existing requirements of the EC GMP Guide and the German Operation Ordinance for Pharmaceutical Entrepreneurs (PharmBetrV).

In addition, the terms of reference of the enterprise (contract manufacture) and the existing structural and sequential organisation were to be taken into account.

On the basis of its activities as a contract manufacturer, reliability, flexibility, assurance of product characteristics and price-worthiness were identified as properties of the organisation relevant to success. These inputs were defined as quality policy and were to be implemented via the Quality System.

A period of one year was envisioned for the execution of the project. The project phases are described in Figure 8.1.

In order to establish a company-wide Quality System, not only the requirements of the EC GMP Guide but also the requirements of ISO 9000 were to be covered. The company chose ISO 9002 as the form for demonstrating its conformance because this model contains all elements with the exception of research and development (R&D), which does not fall within the scope of a contract manufacturer.

Figure 8.1. Timetable for establishing the Quality System.

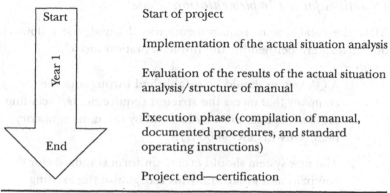

Start   Start of project

               Implementation of the actual situation analysis

Year 1   Evaluation of the results of the actual situation analysis/structure of manual

               Execution phase (compilation of manual, documented procedures, and standard operating instructions)

End   Project end—certification

Appointment of the Quality Representative

After the management had expressly declared its intent to establish a company-wide Quality System, the Quality Representative had to be appointed. This task was performed in the company by the head of Quality Control (QC). An additional job description was drawn up to define and delimit the tasks, responsibilities, and competencies.

The Quality Representative acts as a link between management and the various management levels. On the basis of past experiences, it was considered indispensable to assign a Project Representative who could concentrate solely on the establishment and certification of the Quality System. Both internal and external staff may be used for this. The company decided on one of its own staff in order to use the existing knowledge and experience for the continued maintenance of the Quality System. Under the supervision of the Quality Representative and of the Project Representative, a project team was formed consisting of expert representatives from all areas.

The Implementation Model

Objectives for the Implementation Model

After the catalogue of requirements was defined, the following objectives were defined for the implementation model:

1. A QA system should be established throughout the company that meets the strictest requirements, including regulations laid down by law and by the non-regulatory field (DIN EN ISO 9000 ff).

2. The new system should create uniform standards for the company as a whole and thus reorganise the existing documentation.

3. The system should be established in a process-oriented fashion and serve to ensure that the implementation of high-quality standards is as efficient as possible.

The Process-Oriented Model

Before the process model is explained, it is necessary to explain what is to be included under process orientation from the business management standpoint.

What is Process Orientation? For many years much attention was paid in the description of sequences to the fulfilment of individual tasks. Modern management theories now focus on the process. According to this, the business process can be characterised as a thread running through a company's performance of services, beginning and ending with the external customer. In a functional structure, therefore, the "thread" passes through various hierarchical and functional levels (see Figure 8.2).

Here the customer/buyer defines its parameters regarding costs and time. An efficient implementation across all hierarchical levels must therefore not only define quality, but also minimise the cost and time factors. Every business process can be broken down into manageable main processes (production, QC, pur-

Figure 8.2. Representation of a process sequence through a functionally organised company must be optimised with regards to costs, time and quality.

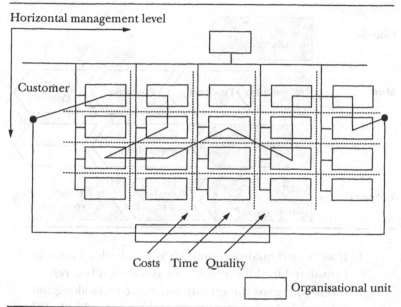

Horizontal management level

Customer

Costs Time Quality

Organisational unit

chasing and so on) for each of which a "process owner" can be identified. In addition, support processes such as personnel and logistics can be identified. In a pharmaceutical company, for instance a contract manufacturer, the business process shown can be broken down into main and part-processes. Figure 8.3 shows the process hierarchy.

Thus the head of Distribution is responsible for his main process and the part-processes resulting therefrom. If the statements made are implemented in a pharmaceutical company, it can be seen why it is advantageous to organise the chapters of the manual according to the structuring of operations. As the result of this method, the process monitor is represented even at, the uppermost document level for his part in the overall process, that is, process optimisation can take place at the highest level.

To understand process orientation, it is important to know that organisational theory is divided into structural and sequential organisation:

Figure 8.3. Example of the division of the business process (contract manufacturing) first into main processes (distribution, purchasing, etc.) and then into part-processes.

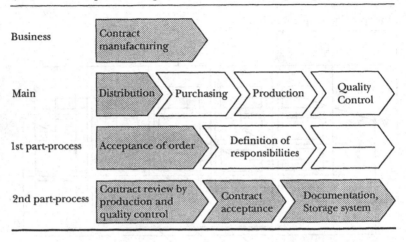

1. If we regard quality assurance as a task which must be considered in the *sequential organisation*, it is surely useful to regard the quality assurance tasks along the process chain and to document them accordingly. What follows from this is that every process owner should be responsible for his own quality-relevant sequences, and therefore for the documentation of these sequences.

2. The *structural organisation* represents the tasks and the responsibilities in the company. The structural organisation is illustrated by the organisation chart. Functional structures are often found in pharmaceutical companies. The organisation chart contains the individual functional divisions of Production, Quality Control, Purchasing and so on (see Figure 8.4).

When establishing a Quality System (in particular in the compilation of the manual), attention must be paid to these structures, since otherwise the heads of the organisational units cannot be clearly assigned to their quality-relevant sequences and documents. In many companies the Quality Manual is organised according to the 20 elements of ISO 9001. How does the head of QC know

Figure 8.4. Example of a functioning organisation.

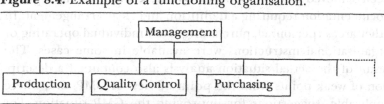

which of the 20 chapters he is completely or partially responsible for? Isn't it better in this case to make a chapter entitled "Quality Control" that describes all his process-related tasks? It is most efficient for the description of the QA tasks to be based on the individual organisation chart (structural organisation).

How Does One Obtain a Process-Oriented Documentation? In order to determine the organisational structure, that is, the structural organisation and the sequential organisation, an analysis of the actual situation is necessary.

Analysis of the Actual Situation

After the project was commissioned on 7 December 1994, the first step to be conducted was the implementation in December 1994 of the actual situation analysis by two auditors at the Jena and Weimar sites. The actual situation analysis served to determine the organisational structure, the existing sequences, the existing documentations, the GMP situation, and the interfaces to be regulated between the various departments, and to clarify the question as to what areas of the company were to be integrated into the Quality System.

Results of the Actual Situation Analysis

The initial situation as regards the documentation was very different at the Jena and Weimar production sites. In the production of parentals at Jena there had already for some years been a "quality assurance ordinance" with controlled documents which already fulfilled many aspects of a QA system. At the newly founded Weimar site, the documentation system had not yet been achieved to this extent. In QC there were various types of documentation requiring a definition and clear arrangement. In other areas (personnel, purchasing, etc.) individual operating or organisational instructions were available in some cases. The result of the actual situation analysis also contained a description of weak points from the point of view of GMP and, where applicable, suggestions for improving the GMP situation. For instance, in QC, standard operating instructions (SOPs) on the handling of reagents, standard solutions or reference standards had to be drawn up or the mode of procedure in the case of out-of-specification results had to be regulated. Responsibility delimitation contracts were also to be concluded with the external laboratories conducting tests on a contract basis.

How is the Quality Manual to be Structured?

Since a Quality System must always be structured from the top down, there is the question of the structure of the Quality Manual as the highest document in a Quality System. According to the catalogue of requirements shown at the outset, the chapters of the EC GMP Guide (requirement no. 1), the 19 chapters of ISO 9002 (requirement no. 2) must be combined in a process-oriented fashion, taking into account the existing organisation and documentation (requirement no. 3). Figure 8.5 illustrates the problem.

As has already been explained, the company organisation (organisation chart) specified important inputs as to what form the structure of the manual should have. If we consider the nine chapters of the EC GMP Guide we notice that the individual chapters can be clearly assigned in the organisation chart to the individual bearers of responsibility, for instance Chapter 6, "Quality Control", to the head of QC, Chapter 3, "Premises and Equipment", to the head of Engineering, and so on. The intersectoral chapters, such as Chapter 9, "Self-inspection", and Chapter 1, "Quality Management", can clearly be assigned to the Quality Representative. From this we can conclude that the EC GMP Guide provides a good basis to which other chapters can be added. "Purchasing and Logistics" (Chapter 10), "Distribution" (Chapter 11) and "Quality Controlling" (Chapter 12) were added in order to be able to present the entire company process in the manual. Figure 8.6 illustrates the formation of the chapters.

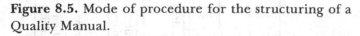

Figure 8.5. Mode of procedure for the structuring of a Quality Manual.

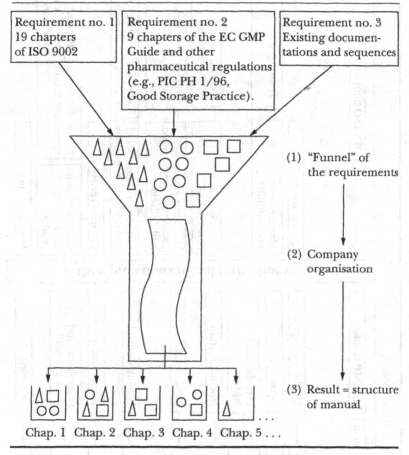

It must be stressed again that the 12 chapters are to fulfill both the requirements of ISO 9000 and those of the EC GMP Guide. Of course, the requirements of the EC GMP Guide are far more detailed in Chapter 5 (Manufacture) and Chapter 6 (QC), and far exceed ISO 9002. The situation is quite the opposite, however, in the case of the chapters "Purchasing and Logistics" and "Distribution", for instance. Here ISO 9002 dominates because the pharmaceutical bodies of regulations define only a few inputs here.

The result of the actual situation analysis was presented at the end of January 1995, in the form of a manual structure which already contained all the structural items of the future Quality

Figure 8.6. Integration of GMP and ISO 9000.

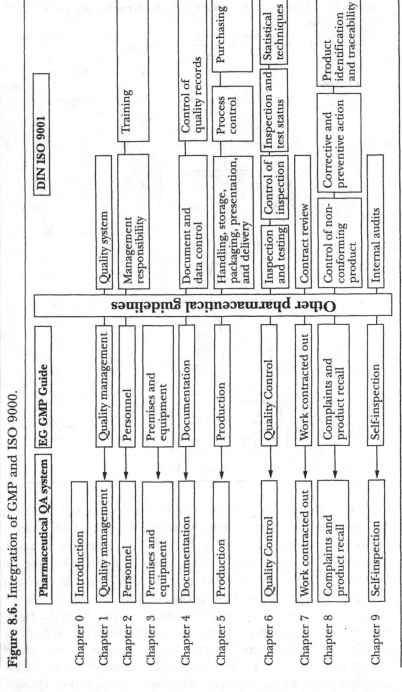

Continued on the next page.

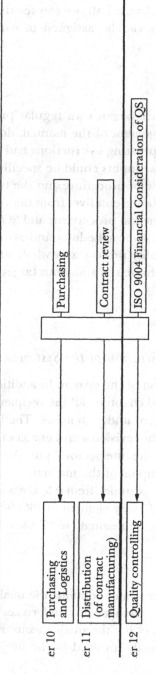

Continued from the previous page.

System. If we transfer the chapters of the manual into the organisation chart of Schering Produktionsgesellschaft, we can see that the corresponding QA responsibility can be assigned to each process owner (Figure 8.7).

Execution Phase

In February 1995, the execution phase began with regular project meetings. First the formal arrangement of the manual, documented procedures and standard operating instructions had to be discussed and defined so that these inputs could be specified uniformly to all areas. Then the mode of procedure and the targets were discussed with the Project Representatives from the various areas. Here the manual, departmental procedures and SOPs were to be compiled simultaneously. Time schedules and action plans were defined both for the quality project as a whole and for the individual areas in order to achieve the ambitious target— certification after 12 months.

Implementation

Chapter 0 Introduction—Administration of the System

Chapter 0 concerns the administration of the system. In addition to the revision status of the individual chapters, all the recipients of a copy of the manual are listed here under Item 0.4. The list of the documented procedures with the revision status enables the reader of the manual an overview of the intersectoral guidelines, which further cement the specified inputs of the manual.

By means of the comparison matrix under Item 0.3, the auditor of the accredited agency can find every element of ISO 9002 and every subitem of this standard in the manual (see Table 8.1).

Chapter 1—Quality System

Chapter 1 describes the Quality System and contains the quality policy of the management. Here the management undertakes to provide sufficient funds and personnel for the quality assurance measures. The management was directly involved in the project

Figure 8.7. Organisation chart of Schering GmbH & Co. Productions KG.

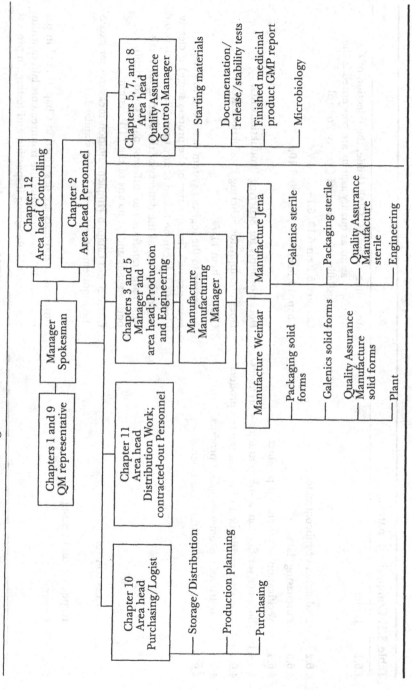

Table 8.1. Comparison matrix.

4.6.1	General	According to the EC GMP Guide, the selection of the suppliers as far as quality aspects are concerned is the responsibility of the head of Quality Control.
4.6.2	Evaluation of subcontractors	Chapter 6 Item 6.13, 6.14
4.6.3	Purchasing data	Chap. 10 Item 10.3 , 10.4 and 10.5
4.6.4	Verification of purchased product	
4.6.4.1	Supplier verification at subcontractor's premises	Does not apply for Schering Produktions KG.
4.6.4.2	Customer verification of subcontracted product	Does not apply for Schering Produktions KG.
4.7	Control of customer-supplied product	Chap. 11 Item 11.7
4.8	Product identification and traceability	According to the logistics system described in Chap. 11 every product (irrespective of the process stage) is clearly identified. It can be traced at any time. The identification begins as early as the delivery of the starting materials, which are transferred to different pallets and identified both visually and by means of the EDP system.
4.9	Process control	Chap. 3 and Chap. 5 [Implementation measures far exceed the requirements set forth in the standards.]
4.10	Inspection and testing	
4.10.1	General	Inspections and tests are described in Chap. 6 (and in part in Chap. 10) far exceeding the requirements laid down in the standards. Receiving inspection and testing, in-process

inspection and testing and final inspection and testing including sampling are described in Items 5.7, 6.6, and 6.7. The content and extent of the inspection and testing activities are defined for each product in the specifications.

4.10.2	Receiving inspection and testing	Chap. 6 (see above) Chap. 10 Item 10.8.2
4.10.2.1	Quarantine status for incoming product not yet inspected	Chap. 6 Item 6.4.4
4.10.2.2	Reduced/restricted receiving inspections and tests	Chap. 6 Item 6.4.3
4.10.2.3	Risk withdrawal	Chap. 6 Item 6.9
4.10.3	In-process inspection and testing	Chap. 5 Item 5.7 Chap. 6 Item 6.4.3
4.10.4	Final inspection and testing	Chap. 6 Item 6.4.3
4.10.5	Inspection and test records	Description in Chap. 6 Item 6.7 The testing record forms part of the batch documentation. Chap. 4 Item 4.8.1
4.11	Control of inspection, measuring, and test equipment	
4.11.1	General	Documented procedures exist separately for QA and production (see below).

right from the beginning. In addition to the implementation of the inputs required by law in § 1a of the PharmBetrV, the efforts to bring about an increase in efficiency in the form of continuous improvement were a decisive argument for establishment of the system. In order to be constantly informed about the current status of the system even after the establishment phase, the "Management review" is a central body of management instruments. Not only ISO 9002 requires this involvement of the management, but also § 1a of the PharmBetrV with its requirement for the active participation of the management. A documented procedure ensures that the Quality Representative prepares a review report at least once a year which contains:

- a summary of the self-inspection;

- a summary of the external inspections (customers, authority);

- recommendations for corrective measures; and

- an overall evaluation of the QA measures taken by the Quality Representative.

Chapter 2—Personnel

Chapter 2 of the manual is compiled by the staff area for personnel. Documented procedures for job descriptions and for determining the training requirement have been drawn up which define the inputs for the managers of the individual areas. In Chapter 2 it was observed that ISO 9002 contains no further-reaching requirements for quality measures. Several useful suggestions for the implementation of QA measures were established from the standard.

Typical deviations to be eliminated in the personnel area by means of the Quality System:

In the areas of manufacture and QC in particular, training courses are currently being held. But it can be said that a systematic implementation, that is, regular determination. of the requirement for training plans, rarely takes place. This also applies to the documented proof of training courses, which is particularly necessary when inspections take place.

Chapter 3—Premises and Equipment

In a pharmaceutical enterprise with a sterile department, the division into relevant clean area grades and the requirements concerning the air take centre stage. Furthermore, the servicing, installation, maintenance and cleaning of the production equipment are important. From the point of view of ISO 9001, the calibration of the measuring equipment used in the production area is the most important topic in this chapter. DIN 10012 [8] provided many suggestions for the implementation of the calibration measures.

Typical deviations in the area of premises and equipment to be eliminated by a Quality System.

It can also be seen that servicing and repairs are conducted. Usually, however, there is no organisational framework containing a relevant documentation or indication of the new maintenance measures to be conducted.

Chapter 4—Documentation

Chapter 4 describes the various document types used at Schering. The four-level model has been developed for the demonstration of the system (see Figure 8.8). It systematises the existing documentation and defines uniform inputs for documents to be compiled anew. Schering was also successful in integrating the existing QA documents into this four-level model.

The system-related documents of Level I contain the Quality Manual with the company-wide quality inputs, the documented procedures containing intersectoral guidelines (e.g., documented procedure for compilation of standard operating instructions or for determination of training requirements), as well as the standard operating instructions containing detailed instructions for the various areas. The manual and documented procedures are

Figure 8.8. The four-level model.

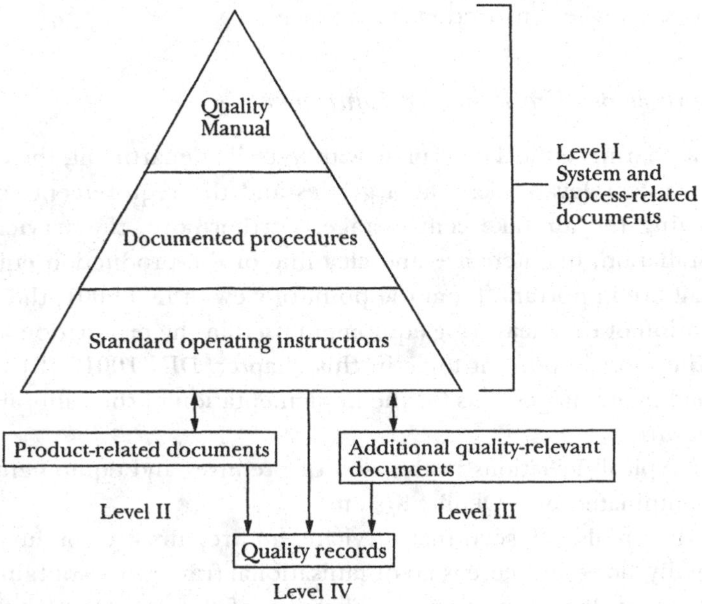

```
                    Quality
                    Manual
                                        ┐ Level I
                                        │ System and
            Documented procedures       │ process-related
                                        │ documents
        Standard operating instructions ┘

  ┌──────────────────────┐      ┌──────────────────────────┐
  │ Product-related documents │  │ Additional quality-relevant │
  └──────────────────────┘      │ documents                │
                                └──────────────────────────┘
        Level II                        Level III

              ┌──────────────────┐
              │ Quality records  │
              └──────────────────┘
                   Level IV
```

managed by the Quality Representative, while the standard operating instructions are the responsibility of the heads of the respective areas, such as Manufacturing, Control or Distribution.

This has the following advantages:

- Detailed regulations are managed, compiled and approved decentrally (according to the inputs in the documented procedure on the compilation and maintenance of SOPs).

- No extensive central QA department need be formed.

- Necessary changes can be implemented rapidly and therefore lead to an enhancement of the flexibility of the overall Quality System.

The area heads are responsible for the GMP-conformant mode of operation in their areas. This also includes the elaboration of SOPs to an adequate extent. The established internal audit/self-inspection system, in connection with the review by the management, regularly monitors the implementation.

At Level II the product-related documents such as manufacturing and test instructions are compiled. Level III quality-relevant documents, which may by no means be neglected, include job descriptions, layout plans of buildings and Site Master Files (SMFs). Level IV comprises such quality records as calibration records and qualification reports. Quality records are regarded separately since they represent a core element of ISO 9002 (Element 16). The inputs for the documents of Levels II, III, and IV are basically fulfilled by documents of Level I, for example, by the standard operating instruction for the compilation of manufacturing instructions which ensure that all items are taken account of in the compilation, administration and modification of a manufacturing instruction. Similar conditions also apply to records, for instance, for calibration. Here attention is to be paid in particular to the content-related inputs of the record and the archiving of the results.

Chapter 4 contains both specified inputs for the EDP system, of which basically the same demands are to be made as of the paper documentation. For instance, if a mask for a test record is stored in the PC, it must *inter alia* be ensured that the modification is protected by password and can only be performed by authorised persons. Therefore, computer validation is a task which is described in Chapter 4.

Chapter 5—Production

In addition to a description of the production processes for the manufacture of solid dosage forms and parenterals, the Quality Manual deals in detail with the following topics: prevention of cross-contamination, prevention of mixups and adulterations and in-process controls, as well as quality planning and process validation.

Chapter 6—Quality Control

Chapter 6 of the manual deals with GMP requirements for the area of QC. With a view to a later ISO certification, in particular the quality elements "Inspection and testing", "Control of inspection, measuring, and test equipment" (meaning calibration of the measuring equipment), "Inspection and testing status" and "Control of nonconforming product" were emphasised here. The evaluation of suppliers is also dealt with in detail in this chapter.

Typical deviations in the area of QC to be eliminated via the Quality System follow.

Qualification/Validation. In QC there are often calibrated pieces of equipment, which are marked accordingly. In the areas of qualification and validation there is still some catching up to be done. For instance, it cannot be assumed that the manufacturing devices have all been structurally qualified according to a systematic qualification in the stages Design Qualification, Installation Qualification, Operational Qualification, and Performance Qualification. In the area of QC, qualification measures have only been taken up on a major scale since the publication by Freeman et al. (1995).

Systematic Evaluation of Suppliers. A functioning evaluation of suppliers presupposes an intense exchange of information between Purchasing and QC and Production. Without a precise description of the interface, such as the tasks and responsibilities defined in a documented procedure, no systematic evaluation of suppliers is possible. This indicates that the inclusion of areas that are not classical GMP areas, such as purchasing and distribution, is necessary. This is generally not the case without a company-wide Quality System.

Chapter 7—Manufacture and Testing on a Contract Basis

Schering Produktions KG operates as a contract manufacturer but also contracts out individual technological steps itself in the production process of certain products to contract manufacturers (e.g., for the manufacture of soft gelatine capsules or sprays). The chapter "Manufacture and testing on a contract basis" refers to this passing on of orders to third parties, and has in particular the contractual delimitation of responsibility as its topic.

Chapter 8—Complaints and Product Recall

Chapter 8 specifies all the statutory requirements from § 14 of the German Operation Ordinance for Pharmaceutical Entrepreneurs (PharmBetrV) regarding complaints for Schering Produktions KG, and indicates the in-company catalogue of measures in the form of a documented procedure.

Chapter 9—Self-Inspection and Internal Audits

ISO 9002 requires regular internal quality audits (focus: proof of maintenance of the functioning capacity of the Quality System); GMP for Medicinal Products expects regular self-inspections (focus: GMP and conformity with marketing authorisation). In order to increase efficiency, self-inspection and internal audit are combined.

Chapter 10—Purchasing and Logistics

The purchasing and ordering system for production materials with article numbers (starting materials, packaging materials, purchased semifinished products, bulk products and finished products) are described in Chapter 10. The purchase of investment goods and services was also defined. This does not include purchasing processes of materials which are not subject to testing or batch-bound, for instance office material.

From the point of view of ISO 9000, the list of admitted suppliers includes not only the manufacturers of raw materials or packaging materials, but also, for instance, a list of the cleaning agents and pesticides used, or a list of the admitted forwarding agents (as far as quality-relevant activities are carried out) including annual evaluation.

The competencies and responsibilities of the logistics department within the area of purchasing and logistics include the storage areas, acceptance of incoming goods and the automatic transport system, as well as commissioning and shipment. All these areas are characterised by a high degree of automation, which makes an extensive description of the storage management systems, the factory process control systems, container transport systems and the SAP system necessary.

Production planning is also located in the area of purchasing and logistics, and is therefore explained and defined in Chapter 10.

Chapter 11—Distribution Contract Manufacture

Distribution Contract Manufacture means the distribution of production capacities (manufacturing output) of Schering Produktions KG and represents an interface between the customers and the production operation. Focal areas of the distribution activities are the issuing of quotations, conclusion of contracts, acceptance and confirmation of orders, coordination of contract manufacture and invoicing.

Chapter 12—Economic Efficiency

A decisive target of the Quality System is the reduction of the costs caused by inadequate quality (see Figure 8.9). Costs of inadequate quality in a pharmaceutical production operation occur in many forms:

- Destruction
- Reworking
- Processing of returned product
- Supply bottlenecks due to blocked release
- Additional administration costs caused by organisational shortcomings

The costs of inadequate quality are only partially covered by the traditional instruments of cost accounting such as cost centre and order accounting. The costs for destruction, processing of returned product and reworking can be identified in the cost

Figure 8.9. Reduction of quality costs.

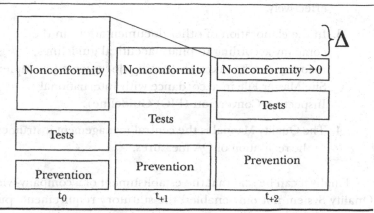

accounting. The additional administrative costs caused by quality defects can at best be determined by means of a operational comparison. In general it is such that the costs of inadequate quality are underestimated in day-to-day operation. The controlling instruments enable the costs of QA and the costs caused by inadequate quality to be shown.

Project Conclusion

Within one year and three months, the existing QA structures at Schering Produktions KG were integrated into a company-wide Quality System. Above all, the circumstance that Schering Produktions KG was founded out of Jenapharm while the system was being developed led to the extension of the planned realisation period by three months.

The establishment of a pharmaceutical Quality System is a complex task. The focus must be on the implementation of the pharmaceutical bodies of regulations.

The comprehensive description in the Quality Manual and other quality documents of the operation sequences in the enterprise has a number of advantages.

1. The Quality Manual gives future guests or inspectors a concentrated, well-structured general overview of the company.

2. New staff can be familiarised with their tasks more effectively.

3. In the elaboration of other documentations in the company according to pharmaceutical guidelines, recourse can be taken to existing documents (example: Site Master File in accordance with International Inspection Convention (PIC) Guideline).

4. The Quality Manual is the central management instrument in the realisation of QA measures.

Finally it can be said that the establishment of a company-wide Quality System not only enabled the statutory requirements pursuant to § 1a of the PharmBetrV to be implemented but that at the same time an instrument for increasing efficiency and customer orientation was created.

Extra Costs Due to ISO 9000? Extra Costs Due to Certification?

After completion of the establishment phase of the Quality System, it can be observed that no extra work resulted from the procedure of assigning the relevant ISO elements to the individual chapters of the EC GMP Guide. In the core areas of production, QC, premises and equipment, the bodies of pharmaceutical regulations are more precise and more comprehensive. There are only few specified inputs for the non-GMP areas, although Chapter 1 of the EC GMP Guide defines a Quality System as having to exceed the GMP areas. The elements of ISO 9000 ff can provide useful hints here. The prerequisite is that one uses the elements of this standard as a checklist and not as a structural characteristic. In the system of Schering Produktions KG three additional chapters were added to the nine chapters of the EC GMP Guide (distribution, purchasing and logistics, economic efficiency). This addition would have been useful and necessary anyway, even without DIN EN 9002.

The certification causes extra costs, of course. A pharmaceutical enterprise does not generally require certification. A leading auditor of TÜV Rheinland formulated it as follows: "Certification is a formal external confirmation that the corresponding standard of series 9000 ff was applied. A certification should only be aimed for if the costs for it can be earned on the market".

For Schering Produktions KG, a pharmaceutical contract manufacturer, the certificate according to the agency accredited pursuant to DIN EN 45012 [10] is an important marketing argument, especially in other European countries with the result that the costs have meanwhile been paid off.

Additional QA measures (e.g., staff training, calibration, qualification) cause higher QA costs. Effective QA measures reduce the costs caused by inadequate quality since they reduce the expenses for destruction, reworking, and so on. Efficient measures lead to reduction of total quality costs.

EXPANSION OF THE QUALITY SYSTEM

A Quality System is never static, since every enterprise is subject to permanent change. Whether this be the result of internal changes (new products, new staff, and so forth) or external changes (new regulatory requirements). Schering Produktions KG received the certificate pursuant to ISO 9002 in March 1996 and shortly thereafter the GMP inspection of the local German GMP supervisory authority took place without complaints. A few months later the company and the Quality System were confronted with two new requirements.

FDA Compliance

As the result of the redistribution of products within the company group, products for the U.S. market also came to the Weimar site. Consequently, a pre-approval inspection by the FDA had to be prepared. One often tends to accord the FDA requirements such a high status that the already existing requirements (EC GMP) appear completely meaningless. In the case of Schering Produktions KG, a systematic elimination of weak points and a structured documentation in accordance with the requirements of the

EC GMP Guide and ISO 9002 had taken place shortly before the FDA requirements were added. Contrary to the frequently expressed opinion that the FDA cGMP requirements far exceed the requirements of the EC GMP Guide, it was observed that the differences are not nearly as large. Adjustments were, of course, necessary in the individual areas. Only the requirements with respect to change control and process validation need be mentioned as examples here. If we look at the Code of Federal Regulations (CFR) we find no statement about a Quality System. The CFR defines the GMP requirements very accurately. The concept of a Quality System within the meaning of ISO 9000 Guide does not exist. Now one can certainly argue about whether a Quality System can be derived indirectly from the CFR, but the CFR contains no concrete system requirements.

Accordingly, the preparations of Schering Produktions KG dealt essentially with the demonstration of conformance at the SOP level. The manual and documented procedures were only adapted to a minor extent with regard to change control and process validation. This minor extent of course only refers to the Quality System. Concerning GMP compliance the difference between EC GMP regulations and FDA requirements was much larger.

The pre-approval inspection by the FDA took place in January 1998. The Quality System was a good instrument for this in order to demonstrate the implementation of the GMP requirements from the upper management level (Quality Manual) via the middle management level (documented procedures) to the personnel level (SOPs). The result of the pre-approval inspection: only minor deviations—no 483s!

Environmental Management

In addition to the quality requirements, the environmental protection requirements are another important aspect in which an enterprise must demonstrate its conformance. In Europe, the regulation 1836/93, known as the "Eco Audit Regulation", defines the inputs. At the global level ISO 14001 "Environmental Management Systems—Specification with guidance for use" is the standard for an environmental management system (see Table 8.2).

Table 8.2. Content of ISO 14001.

Introduction	
1	**Scope**
2	**Normative references**
3	**Definitions**
3.1	Continual improvement
3.2	Environment
3.3	Environmental aspect
3.4	Environmental impact
3.5	Environmental management system
3.7	Environmental objective
3.8	Environmental performance
3.9	Environmental policy
3.10	Environmental target
3.11	Interested party
3.12	Organisation
3.13	Prevention of pollution
4	**Environmental management system requirements**
4.1	General requirements
4.2	Environmental policy
4.3	Planning
4.3.1	Environmental aspects
4.3.2	Legal and other requirements
4.3.3	Objectives and targets
4.3.4	Environmental management programme(s)
4.4	Implementation and operation
4.4.1	Structure and responsibility
4.4.2	Training, awareness and competence
4.4.3	Communication
4.4.4	Environmental management system documentation
4.4.5	Document control
4.4.6	Operational control
4.4.7	Emergency preparedness and response

Continued on the next page.

There is no doubt that the combination of quality and environmental management systems is the trend for the future. ISO 14001, for instance, contains a matrix which makes the relevant connection to ISO 9000.

It is also the declared objective of ISO to further harmonise the two standards. The combination of the two systems also makes sense in the field of the pharmaceutical industry. This can be illustrated by the following example.

Cytostatics are produced in a laminar flow bank. From the point of view of GMP, it is important that the filter in the LF bank be changed regularly and that this be documented. As a rule, there will be an SOP for this process. The SOP often ends with the installation of a new filter. From the point of view of environmental protection it is, however, decisive how the contaminated filter is disposed of. This means that it must be clear to the employees where to dispose of the filter. The classification of the filter in a waste class and the documentation of the disposal route including handing over the filter to a disposal company which has the relevant authorisation must be regulated. Here a process chain becomes clear which should in any case be described in an SOP. A quality SOP and an environmental SOP will surely not make the work of the employee on site easier. This process should be described in *one* SOP.

As has already been described, Schering Produktions KG has an extensive Quality System that was to be expanded to include the requirements of ISO 14000 and the EC Eco Audit Regulation. The described Quality System with the 12 chapters of the manual was expanded to include Chapter 13 "Environmental management system" (see Table 8.3).

Chapter 13 of the manual regulates the system inputs for the environmental management functions. It is therefore the counterpart to Chapter 1 "Quality System", which regulates the system inputs for the quality management functions. Figure 8.10 illustrates the new structure.

After the addition of Chapter 13 to the manual, Chapters 2 to 12 were supplemented with the corresponding environment-specific inputs.

In Chapter 3 "Premises and equipment", the item "Environmental protection measures within the framework of the use of premises and equipment in production" was added. It describes, for instance, how waste from production plants should be dealt with, such as contaminated filters, and cleaning and lubrication agents, as well as how waste water as in the case of cleaning processes in the plants can be reduced. This chapter of the manual refers to concrete documented procedures, "Collection, delivery and control of waste and recyclable materials", which are valid for the entire company.

As the result of this expansion of the individual chapters of the manual to include environment-specific requirements, the principle was upheld that each process owner is responsible for their own quality- and environmentally relevant tasks and documentation.

In December 1998, Schering GmbH & Co. Produktions KG was issued the certificate pursuant to ISO 14001, one month later the environmental declaration was validated in accordance with EC Regulation 1836/93.

CONCLUSION

The trend nowadays is clearly toward process-oriented management systems. In the pharmaceutical companies many quality requirements [EC GMP, cGMP (FDA)] and other requirements

Table 8.3. Correspondence between ISO 9001 and ISO 14001.

ISO 9001 (1994)		ISO 14001 (1996)	
Management responsibility			
Quality policy	4.1.1	4.2	Environmental policy
	—	4.3.1	Environmental aspects
	—[1]	4.3.2	Legal and other requirements
	—[2]	4.3.3	Objectives and targets
	—	4.3.4	Environmental management programme(s)
Organisation	4.1.2	4.4.1	Structure and responsibility
Management review	4.1.3	4.6	Management review
Quality system			
General	4.2.1 1st sentence	4.1	General requirements
	4.2.1 without 1st sentence	4.4.4	Environmental management system documentation
Quality system procedures	4.2.2	4.4.6	Operational control
Quality planning	4.2.3	—	
Contract review	4.3[3]	4.4.6	Operational control

Continued on the next page.

Continued from the previous page.

Design control	4.4	4.4.6	Operational control
Document and data control	4.5	4.4.5	Document control
Purchasing	4.6	4.4.6	Operational control
Control of customer-supplied product	4.7	4.4.6	Operational control
Product identification and traceability	4.8	—	
Process control	4.9	4.4.6	Operational control
Inspection and testing	4.10	4.5.1 1st and 3rd paragraph	Monitoring and measurement
Control of inspection, measuring and test equipment	4.11	4.5.1 2nd paragraph	Monitoring and measurement
Inspection and test status	4.12	—	
Control of nonconforming product	4.13	4.5.2 1st part of 1st sentence	Nonconformance and corrective and preventive action
Corrective and preventive action	4.14	4.5.2 without 1st part of 1st sentence	Nonconformance and corrective and preventive action
	—	4.4.7	Emergency preparedness and response
Handling, storage, packaging, preservation and delivery	4.15	4.4.6	Operational control
Control of quality records	4.16	4.5.3	Records

Continued on the next page.

Continued from the previous page.

Internal quality audits	4.17	4.5.4	Environmental management system audit
Training	4.18	4.4.2	Training awareness and competence
Servicing	4.19	4.4.6	Operational control
Statistical techniques	4.20	—	
	—	4.4.3	Communication

[1]Legal requirements addressed in ISO 9001, 4.4.4

[2]Objectives addressed in ISO 9001, 4.1.1

[3]Communication with the quality stakeholders (customers)

Figure 8.10. Expanded structure of the quality and environmental management manual.

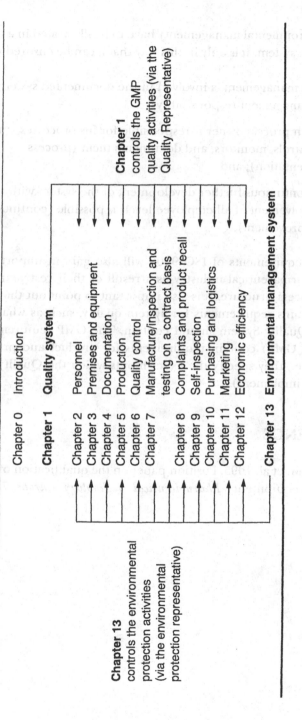

(e.g., environmental management) have to be illustrated in a comprehensive system. It is only in this way that it can be ensured that:

- the management is involved in the documented system (management responsibility);

- each process owner is responsible for his processes, and controls, monitors, and documents them (process orientation); and

- a continuous further development of the system with involvement of all employee levels is possible (continuous improvement).

The requirements of ISO 9000 will also gain in importance in the pharmaceutical industry as a result of their restructuring in the process structure. It is also important to point out that ISO 9000 contains requirements for system quality, such as what elements a Quality System should contain. The GMP requirements (CFR, EC GMP) contain requirements for product and process quality (i.e., they specify the exact inputs which the Quality System must implement).

REFERENCE

Freeman, M. et al. 1995. Position paper on the qualification of analytical equipment. *Pharmaceutical Technology Europe* 7(10): 40–46.

9

Experiences in the Auditing of Quality Assurance Systems: Industry's View

Andreas Brutsche
Novartis Pharma AG
Basel, Switzerland

QUALITY MANAGEMENT—IMPLICATIONS FOR THE PHARMACEUTICAL INDUSTRY

Managing quality has been going on for centuries, in a good, bad, or indifferent way. In recent years this process has become known as Total Quality Management (TMQ). What is most important is the *commitment* of the whole business to the idea that quality (work, process, product and so on) is organized in a *systematic* manner. Equally important is that the employees understand and share this commitment on a long-term basis.

Quality is an attitude, and is closely associated with how people behave, how they perform, and what training they have received. It also has to do with leadership and organization—in other words, with what priority is awarded to quality in all functional groups within the company.

Quality Management is a permanent strategy involving every person in every department (research, quality control [QC], production, marketing, purchasing and so on).

The purpose of TMQ is to provide a structured and systematic approach to follow up and continuously develop the quality improvement work throughout a company.

Quality Assurance (QA)

QA has been defined as integrated management systems which provide an assurance that the contractual and legal obligations of the company, to its customers and the community, are being efficaciously fulfilled.

Thus, industry has developed various ways of fulfilling its obligations and as a result the principles of Quality Management, now largely embraced in ISO 9000, have emerged.

However, the adoption of these principles, the establishment of a system of QA, and the creation of management structures necessary to cope with the application of QA has happened and in different ways. In the pharmaceutical industry, many organizations still believe in the view that quality means Good Manufacturing Practice (GMP) (see Table 9.1).

The acceptance of GMP guidelines for pharmaceutical products can be said to be almost universal, and to a large extent companies comply with the principles expressed by such guides. However, the route to achieving compliance may vary considerably from company to company. Much will depend upon how various functions (e.g., production, QC, warehousing, engineering, distribution and marketing) involved in making medicines view GMP and what priority they award to it. In the production and laboratory departments GMP may have a high priority, whereas finance and personnel may have difficulty in recognizing how GMP regulations should be applied to their own activities.

It should nowadays be very clear that GMP is only one element in the structure that determines the total quality of pharmaceutical products. Therefore, the way in which a company manages

---

Table 9.1. Perception of quality management.

1900 Manufacturing Control
 1930 Analytical Control
  1960 Quality Control
   1970 Good Manufacturing Practice
    1977 Quality Assurance
     ? Quality Management

---

the quality of its products should constitute an important aspect of its overall approach.

What Are the Potential Advantages of Quality Management?

The attainment of true Quality Management within a manufacturing operation benefits delivery performance by eliminating wasteful practices, increasing productivity, and providing a higher level of assurance of quality. The rewards of this approach are greater operational efficiencies, lower costs, and improved reputation and competitiveness.

Examples of opportunities in a pharmaceutical manufacturing operation include the following:

- Reduced testing and inspection of incoming materials

- Better use of validation data

- Extension the concept of parametric release

- Improved in-process controls

- Elimination of QC inspection

- Review of end-product testing

All of this means adopting a multifunctional approach to Quality Management, which will result in improved manufacturing efficiencies together with a known quality level in compliance with the company's own quality objectives.

Extrinsic and Intrinsic Quality

Intrinsic Quality. The intrinsic quality of a pharmaceutical product is what the Orange Guide describes. It is its aptness for use, its efficacy, its purity, its exact dosage, and its stability.

Extrinsic Quality. If a tablet has a slight black spot on it, which may come from a chemically detectable impurity in one of its starting materials such as starch, it will in no way affect the safety or efficacy of the tablet. This is an extrinsic quality (or non-quality) feature. If the date of expiry on a carton is not as clear as it should be, or the color of a carton is not quite as green as it should be, these are extrinsic quality problems. They will not affect the safety of the patient. They may harm the image or reputation of the firm who makes it, but the patient will still get full benefit from the product.

Quality is a Competitive Weapon

In the fifties, Japanese cars, motorcycles and cameras were the laughing stock of the West. Three cameras out of four failed after the second roll of film. Motorcycles broke down after 50 miles. Japanese culture cannot stand ridicule, or "losing face", and we all know the result today. But it took 20 years of hard work and investment in development, people, methods and equipment for Japan to arrive at the high quality level that is now being achieved.

Quality in Japan is part of work. I recommend the reader consult many other works which treat this subject in depth.

Quality is and will become, more and more a competitive weapon. I am not referring only to the product, but also to the total quality of the manufacturing and technical operation, be it design, development, manufacture, distribution, logistics, purchasing or stock control.

Quality does not refer only to mechanical or physical work. This message has not yet permeated all companies. It must permeate administration, service, teaching, training and human relations. In the past, the talk was about "efficiency", which is similar but tends to have a cold and impersonal connotation.

Often, quality is not costly, yet it has value!

Assessment of Quality Systems

Auditing is a prime element within the quality management system, but it is important to be clear on terminology.

The objective of a system audit is to examine management's provision for, and commitment to, quality. This is quite different from a compliance audit, which looks at operator understanding and commitment to quality through the manner in which compliance with the system is achieved.

Thus management structures will need to be considered. Key questions include the following:

- How are the personnel organized?

- How is responsibility for the product quality shared?

Some examples of principal components of auditing the quality will include:

- Documentation

- Release procedures (for materials and products)

- System review mechanisms

- Contract agreements

- Regulatory compliance

In addition, there are other aspects which need to be considered, including:

- Complaints procedures

- Recall procedures

- Distribution and returned goods

- Engineering calibration and maintenance

- Technology transfer

- Validation policies and practices

A key feature of this audit is to break the system down into components or units. Thus each unit can be seen to have an input, an operation (or process), and an output. Each aspect will need to be fully satisfied when applying QA to the unit concept.

Thus the auditor must ensure that:

- Each unit is adequately controlled.

- Control systems are adequately described, documented and reviewed.

- Interface arrangements between units are satisfactory.

In the audit of any Quality System, it is of overriding importance to ensure prompt and efficient processing and evaluation of "error messages". It is when the unexpected happens that product quality is at its most vulnerable.

There is no substitute for the application of a well-structured, properly formulated audit program carried out by skilled, experienced personnel to assess and further the QA system.

Those who carry out audits must therefore do the following:

- Establish that a Quality System exists and is complied with.

- Find out if quality is understood at all levels.

- Assess risk to the product and the company.

- Challenge established practices.

The importance of a well-designed Quality Management System should be self-evident. It can, if properly planned and implemented, form the major part of a business strategy, as well as providing confidence that the company's legal and contractual obligations are being fulfilled.

Its main benefit, however, is to improve the company's competitive edge. It should therefore be designed with the following in mind:

- Take every advantage of technology.

- Don't forget about materials.

- Eliminate wasteful practices.

Above all, its importance to the factory management team will be to improve operational efficiency without any sacrifice in quality.

Comparison of FDA and EU Regulatory Inspections

The European Commission (EC) Directive 89/341 provides for the basis of future inspection in the Community. In particular, it provides the legal basis for compulsory compliance with GMP in all member states.

Compliance with GMP will be assessed by member state inspectors who will conduct regular inspections of pharmaceutical companies. They will also ensure that all the legal requirements in member states are in compliance. Of particular importance in the EC will be:

- compliance with the marketing authorization; and

- compliance with the manufacturing authorization.

The directive also provides for the contents of the inspector's reports to be communicated to the manufacturer.

Typical Inspection Program

The regulatory inspector, when performing an official inspection, will have several objectives, as follows:

- Check details of the "authorizations" which may have been granted.

- Ensure compliance with national laws, and enforce if necessary.

- Assess adherence to principles of GMP, and the ability and competence of the company to make products to consistent quality standard.

- Notify the company of any failures or omissions in GMP and seek remedial measures.

The inspector may also need to investigate any complaints or defects which may have been reported, and in addition may take samples of company products for analysis in a government laboratory.

Overview of Differences Between EC and U.S. (FDA) Approaches

To fully compare and contrast the EC and the Food and Drug Administration (FDA) would take a long time. What follows, therefore, is a short summary of the present position.

Legal Differences. In essence, the legal bases are rather similar from the inspection point of view. However, one major difference is the U.S. Freedom of Information Act, which allows the public access to the FDA inspection reports. This is not the case in most EC member states.

Philosophical Differences. The U.S. view is adamant on compliance and enforcement. The EC, so far, tends to operate on the basis of trust and cooperation.

Operational Differences. The FDA is a much bigger organization than the EC equivalents, with very wide-ranging activities covering food and cosmetics as well as drugs. The FDA is much more involved in biologicals and bulk chemicals. The FDA inspectors are much more likely to carry out unannounced inspections in the United States and will also put more emphasis on preapproval inspection.

Technical Differences. In general terms, there are few technical differences, although some specific aspects, such as investigation and handling, are given more emphasis by the FDA—as is the whole subject of validation.

The Role of Auditing and Self-Inspection

Pharmaceutical auditing is a very extensive exercise aimed specifically at assessing the company's QA system. It is "a systematic and independent examination of the effectiveness of the quality system or of its parts". It is important to be clear on the terminology, since the terms "auditing", "investigation" and "self-inspection" can sometimes be used to describe quite different activities.

An *audit* has the primary purpose of carrying out a regular and thorough review of all aspects of an operation and their effect upon product quality. The results of audits should be for senior management to act upon.

Investigations are associated with specific problems; for example, in response to a known product defect. They will normally be confined to certain aspects of the operation where the problem is thought to have arisen, but they can be unpredictable depending upon the findings which arise.

Self-inspections are limited in scope and more mechanical in operation. They are internal. They should probably be carried out more frequently than audits, and the results should be dealt with by local management.

The Justification for Auditing

The products of the pharmaceutical industry are normally the result of a long interrelated sequence of events, from original research and development (R&D), through processing and packaging. There are often:

- a wide variety of products of different therapeutic action;

- many different dosage forms;

- many discrete or "batch" processes with consequent changes;

- many markets to serve; and

- many types of packaging.

Consequently, there are many opportunities for error, the results of which may have serious repercussions not only for the patient, but for the company and its reputation.

In addition, the industry is heavily regulated and likely to remain so in the future. Therefore, one of its tasks is to design and implement an effective QA System which will prevent errors from occurring. It also has a responsibility to review the system periodically, in order to ensure that it continues to be managed effectively. This is the prime reason for the audit and thus becomes an integral part of the system itself. However, there are a number of additional reasons for auditing which should be borne in mind:

- Audits can be used to ensure compliance with regulations (compliance audit).

- Audits can be used to build confidence in the QA System (system audit).

- Audits can provide a basis for mutual trust, technical understanding, and good communication between different departments and disciplines.

- Audits help to promote confidence and good relations between the industry, its suppliers and its customers.

- Audits assist in establishing corporate quality policies on an international scale.

Where and When Are Audits Performed?

It is self-evident that the company's own facilities and QA Systems should be audited, but an audit program should also look at the following areas:

- suppliers of materials;

- suppliers of equipment;

- outside contractors of laboratory work, processing, packaging and printing;

- storage, distribution and shipping; and

Auditing should be carried out on a regular basis as a means of continual assessment, but there may be special reasons which prompt an audit, such as:

- new product introduction;

- new technology;

- new or upgraded facilities; and

- response to regulations/regulatory action.

The Auditor

The success of an audit ultimately depends upon the abilities of the auditor. The ideal characteristics of a pharmaceutical auditor can be summarized as follows:

1. Appropriate technical and professional qualifications.

2. Wide experience in the development, design, formulation, processing, packaging and control of a variety of pharmaceutical dosage forms and active pharmaceutical ingredients.

3. Exposure to the interrelationships of the various parts of a manufacturing organization.

4. Extensive well training in auditing.

5. Possession of a wide range of personnel skills.

Of the attributes mentioned, experience, training and personnel skills are the most important.

Planning and Management of Audits

The plan of campaign for a successful audit and self-inspection program can be usefully broken down into various stages, including the preparation, the audit itself, and the followup.

However, before any of these are undertaken, the audit philosophy must be agreed upon within a company and, where applicable, with suppliers, contractors and so forth. A company should not approach auditing and self-inspection as something that has been imposed upon them by the regulatory authorities, although these requirements must not be ignored. Each company and department/functional group within it should devise a plan of campaign based upon the operations being carried out, their complexity, the environmental effects, processing and so on.

Preparation. Before embarking on an audit program, those who are going to carry out the audits need to be identified. The audit team must be selected and trained, and it must be determined whether they will perform auditing duties full or part time, and who will bear the group's responsibilities.

The whole program, as well as each individual audit, must be carefully managed—with emphasis on ensuring that each one is accepted by the auditee as part of the quality system. Once this is achieved, audits can be carried out with full cooperation and regarded as constructive. A very important component of this overall management is information gathering, which includes the following:

- knowledge of area, site or company to be audited;

- authorities and standards;

- previous audit report findings;

- product(s) in question—"quality" history; and

- complaints, recalls, reworks, deviations and out-of-specification (OOS) results.

Information gathering enables the auditor to decide how the audit will be carried out and what special emphasis or focus should be given to specific parts of a facility or process.

The execution of an audit should always follow a planned format, so time is allowed to cover all required areas in a professional but amicable manner. A suggested format has these steps:

- Opening meeting

- Actual audit

- Daily review

- Preparation for final summary

- Summary session

- Report and follow-up

The opening meeting serves to:

- introduce audit team;
- explain purpose and objectives of audit;
- meet escorts; and
- propose and agree upon program timing.

The audit itself should:

- be coordinated, not confused;
- keep control to ensure progress; and
- obtain copies of documents and samples as required.

A daily review will:

- summarize progress;
- confirm outstanding items and program details;
- ensure that "requests" are processed;
- review any actions taken; and
- discuss any misunderstandings or disagreements.

The preparation for the final summary:

- allows sufficient time;
- utilizes a written structure;
- deals in facts, not opinions;
- relates comments to specific items;
- summarizes under headings;
- draws priorities;
- agrees upon conclusions; and
- anticipates problems.

The summary session:

- explains how it is to be run;
- agrees on facts; and
- offers advice and support.

The following criteria should be included for the report:

- Write the report as soon as possible.
- Create a permanent record of activities and findings.
- Follow a structure.
- Ensure accurate reflection of all proceedings.

Auditors have an obligation to ensure that the report represents the "sampling exercise" undertaken—don't jump to conclusions if all the facts are not known.

Follow-Up. Audit and self-inspection programs often stumble in their follow-up efforts: the audit is complete, the report is written, and then it is filed and forgotten. A system is required for someone to take overall responsibility to ensure that the necessary actions are being furthered and responsibility has been delegated. Depending upon the urgency, senior management will need to be aware of, involved in, and approve the actions proposed. For longer-term projects (GMP upgrade projects), a plan is strongly recommended which clearly states the following:

- What is required.
- When it will be achieved.
- Who is responsible.

The overall program timing should be such that there is ample opportunity for deficiencies to be dealt with before the next audit. Otherwise the auditee may become increasingly frustrated when outstanding action points are carried over from previous audit reports.

Conclusions. Adoption of a proper plan for audits is the key to success. A good audit plan will do the following:

- Establish the auditor as professional, in control of the situation.

- Shortens the audit, thereby minimizing any disruption to the company whilst making maximum use of both the auditors' and the auditees' time.

- Enable the company to plan ahead, thus making the visit as smooth as possible for all those involved.

- Prevent the audit from drifting aimlessly by identifying and focusing attention on the matters to be considered.

- Demonstrate a courteous and interested attitude by taking the time to become familiar with the organization, its personnel and its activities.

- Introduce the role of auditing and self-inspection and its acceptance as part of QA and general management of quality.

Audit Rating Systems. Some firms use audit rating systems, either by giving a number of points for each inspection report or by giving a grade.

Audit rating systems tend to lack objectivity, because the problems are difficult to quantify. An experienced team, however, can give a useful overall grade.

Audit Techniques. There are various approaches which an auditor will take in different parts of the organization. However, in all cases the audit is a sampling exercise, and it is the responsibility of the auditor to assess the entire operation on the basis of the sample taken. The sample must, therefore, be as representative as possible and must include all parts of the organization normally associated with the day-to-day activities of the site being inspected. This can be achieved per the following audits.

Trace Forward Audit. This is the conventional approach, which begins in the warehouse with starting materials, and logically follows the process through dispensing, preparation, manufacture, filling, packaging, labeling, release and dispatch. From this main flow, side activities can be pursued systematically as they are reached.

Trace Back Audit. This type of audit is essentially the reverse of a trace forward audit.

Random. Such an audit involves visiting all elements of the organization in a random order. Although it is a flexible approach, it can fail to highlight cases of poor interdepartmental communication or cooperation.

The Document Trail. This is a method favoured by American auditors, in which all documents relating to a given batch of product are examined and cross-referenced to each other. This will include not only primary batch records and testing results, but also all associated Standard Operating Procedures (SOP), log books, calibration records, transfer dockets, and so on, which may not be part of the batch protocol. The document trail can be a useful method to gain an overview of the QA approach, but usually only highlights cases of simple human error.

To increase the value of audits, all the above methods could be applied after reference to the company's complaint file, or in response to specific defect or recall events.

Product Audit. This may be considered to be a specific case of the trace back audit. The audit starts with a given product and batch number, and the objective is to explore backward through the various activities.

Main Topics which Should Be Covered during an Audit of a QA System

Quality System. Planning for quality involves definition of the quality system and the required quality and GMP compliance standards, in the form of a Quality Manual.

Quality Manuals provide an overview of the quality system and are used for the following purposes:

- communicating the company's policies, procedures and requirements;

- implementing an effective quality system;

- providing improved control of practices and facilitating QA activities;

- providing the documented base for auditing quality systems;

- providing continuity of the quality system requirements during changing circumstances; and

- presenting the quality system for external purposes, such as regulatory inspections.

The separation of QA from production and logistics must be ensured to avoid conflict of interest situations.

SOPs. SOPs describe the detailed working practices, for use by the personnel carrying out the work. The working practices must be in accordance with the requirements of the Quality Manual, and be within the scope permitted by the relevant international policies and guidelines.

- SOPs must be routinely available at the workplace.

- Quality Manuals may be used as checklists to ensure that each mandatory requirement is covered by a corresponding SOP, thereby ensuring that the SOP list is complete.

- The structure and format for SOPs should be adapted to the requirements of the workplace.

Change Control System. Formal change control is an essential feature of any QA System, to ensure that all changes are fully evaluated for their effect on product quality or validated status, and authorized before their implementation. Change control systems also ensure coordination of changes with all parts of the organization and, where relevant, with regulatory authorities.

Contracts for Manufacture and Analysis. Contract manufacture and analysis must be correctly defined, agreed upon and controlled in order to avoid misunderstandings which could result in products or work of unsatisfactory quality. There must be a written contract between the contract giver and the contract acceptor which clearly establishes the duties of each party. The contract must clearly state the way in which the authorized person releasing each batch of product for sale exercises his or her full responsibility.

Documentation and Records Control. Good documentation constitutes an essential part of the QA System. Clearly written documentation prevents errors that arise from spoken communication and permits tracing of batch history. Specifications, master formulae and instructions, procedures, and records must be available in writing, in a language readily understood by the personnel using them.

Documents must be designed, prepared, reviewed and distributed with care. They must comply with the relevant parts of the manufacturing and marketing registration dossier. Documents must be approved, signed and dated by appropriate, competent and authorized person.

Supplier Qualification. A supplier qualification and monitoring system should be in place. The purchase of materials and products is an important operation which must involve staff who have a particular and thorough knowledge of the suppliers. Materials and products must only be purchased from suppliers approved by QA. Whenever possible the material must be purchased directly from the producer. All materials and products purchased must have a mutually confirmed supply agreement/purchase order.

Product Identification and Traceability. A system should be in place for identification and traceability of any component used in the manufacture of a product, including standard nomenclature, code systems, lot numbering, sample identification, reconciliation and distribution records.

Validation. Validation is a complex and highly technical subject which essentially concerns verifying that the process or methodology used for making and packing a medicament is reliable. For this, a number of critical parameters and their influences on the product have to be assessed to demonstrate that the ones chosen give the desired result.

Unfortunately, validation is not a panacea for avoiding errors.

Validation Master Plan. A Validation Master Plan should be available, showing the elements and extent of the qualification and validation program.

Process Validation. Introduction of new or changing processes into manufacturing or must undergo prospective or concurrent validation. Well-established processes can be validated retrospectively through review of historic data and information. An authorized validation protocol should be available prior to conducting the validation exercise, and should include the criteria by which acceptance of the validation study will be measured. For prospective validation, a minimum of three consecutive batches must be successfully manufactured in compliance with specifications and the requirements of the protocol in order to demonstrate that the process is valid. Critical processing parameters must be identified and the process validated within a specified range of operation. A validation report which includes a summary of the data, a reference to the raw data, a statement on the equivalence with the bio-batches and the outcome of the validation study must be written, for review and approval by the validation team and by the QA department.

Cleaning Validation. The primary benefit usually attributed to a cleaning validation is ensuring compliance with federal regulations. However, a more important benefit of cleaning validation work is the identification and correction of potential problems, previously unsuspected, which could compromise the safety, efficacy or quality of subsequent batches of drug product produced with the equipment. It is important to keep the true purpose of cleaning validation studies in mind throughout the planning and execution stages of the project, in order to utilize the valuable resources dedicated to the effort as efficiently as possible. Several serious problems can be prevented through the use of a reasonable cleaning validation program. All are related to adulteration, therapeutic safety and efficacy, or overall quality of the product over its shelf life:

- Cross-contamination with active ingredients

- Contamination with unintended materials or compounds

- Microbiological contamination

Computer Validation. The need for validation applies to computerized systems as much as any other process equipment or facility, and the principles of validation can be equally applied to computers. The process of validation from the earliest stage of a system design provides an audit trial for the history of the system and a constant integrated record which allows for effective management control. A validation of software after the life-cycle concept is a basic requirement for each computerized system used in a GMP-related area (see Figure 9.1). All systems which handle data or processes which fall under the justification of GMP must be validated. Validated systems shall have a validation protocol, defining the planned activities and documentation required for validation, and a validation report summarizing the results of the execution of the protocol and conclusions reached. The protocol must be approved as early as possible, and before operational qualification (OQ)/performance qualification (PQ) is executed. Approval of both protocol and report shall constitute acceptance that the system is validated. Other documentation must be included or referenced in the protocol and/or report. Such reference documentation must be readily accessible should the system be inspected or audited. These documents include user requirements, functional specifications, design specifications, vendor evaluation, system construction documentation, testing plan, results, and reports, validation SOPs, operational SOPs, system log books, user manual and technical documentation.

Batch Record Review and Release Decision. The QA department should review data and/or records associated with the manufacture and control of drug substance and pharmaceutical products prior to any release decision. The records and/or data may be in written, electronic or photographic form.

Control of Nonconforming Product. A system should be in place describing the general requirements for controlling nonconforming product, including investigation, reprocessing or reworking and corrective action.

Customer Complaints. The requirements for the receipt of customer complaints on the company's products, and for subsequent required actions—processing, trending, failure investigation and corrective action, should be described.

Product Recall Procedure. A system for the organization and execution of product recalls should be in place. The requirements for warehouse and distribution records, product reconciliation, product retrieval and storage, documentation and reporting, procedure maintenance and testing of a recall system should be included.

System for Corrective and Preventive Action. There must be procedures in place for the effective handling of product nonconformities. Procedures must include an investigation into the cause of nonconformities relating to product, process and quality system and the recording of the results of the investigation. Investigations must wherever possible include a determination of the corrective action needed to eliminate the cause of the nonconformities.

Records. A policy should be in place that defines the requirements for records in providing a history of each batch of product, including its distribution, and also for validation, equipment maintenance, and cleaning pertinent to the quality of the final product. The different types of records, records media, security, confidentiality and availability for regulatory authorities should be described in such a policy.

Quality Audits. Auditing activities must be described in an approved procedure which defines the objectives, responsibilities, scope, frequency and preparation of an audit program, reports, including followup and corrective action plans, restrictions in circulation and the need for maintaining records. There must be an audit program which identifies the date, place and reason for each audit together with the requirements of an audit team

Figure 9.1. Life cycle for the development of a computer system.

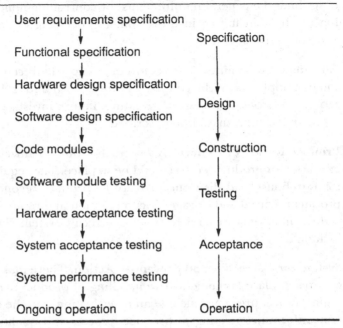

comprised of members with specialist skills as necessary. Internal self-inspections of each departmental activity which has a bearing upon product quality must be carried out on at least an annual basis. These inspections must cover shift activities as necessary. Audits of third-party manufacturing and packaging operations must be carried out on a frequent basis.

Qualification and Training of Personnel. A SOP should be available describing the requirements for carrying out qualification, management, and technical training of personnel. Selection of personnel, job descriptions, appraisals, records, motivation, organization, reviews and training programs should be included.

Trending/Annual Product Reviews. Procedures must be in place for the regular review of all relevant information such as audit reports, quality records, customer complaints and so forth, in order to determine the steps needed to deal with any problems, or potential problems, requiring preventive action.

REFERENCES

Chapman, K. G. 1991. A history of validation in the United States, Part I. *Pharm. Tech.* (November).

Code of Federal Regulations 21. *Current Good Manufacturing Practice for finished pharmaceuticals.*

Crosby, P. B. 1996. *Quality is free.* New York: McGraw-Hill.

Crosby, P. B. 1995. *Quality without tears.* New York: McGraw-Hill.

Deming, S. N. 1988. Quality by design, *CHEMTECH* (September).

DIN EN ISO 9000-1, 9001, 9002, and 9004-1. August 1994. Berlin, Germany: Beuth Verlag.

Goldberger, F. 1991. *Pharmaceutical manufacturing: Quality management in the industry.* France: Herissey Press. ISBN 2-9506062-0-2.

Good Manufacturing Practices for pharmaceutical products (WHO).

Harder, S. W. 1984. The validation of cleaning procedures. *Pharmaceut. Technol.* (May).

Hayes, Wheelwright and Clark. 1988. *Dynamic manufacturing.* New York: The Free Press.

International conference on harmonization of technical requirements for registration of pharmaceuticals for human use.

Ishikawa, K. 1985. *What is Total Quality Control?: The Japanese Way.* Englewood Cliffs, NJ: Prentice-Hall.

Kuzel, N. R. 1985. Fundamentals of computer system validation and documentation in the pharmaceutical industry. *Pharm. Tech.* (September).

The rules governing medicinal products in the European Community.

10

How to Implement Validation into a QA System

Fritz Demmer
Pharmaceutical Consultancy
Hirschberg, Germany

Validation evolved in the United States during the 1970s when large-volume parenterals (LVP) had to be recalled because of product contamination and/or lack of sterility. It became evident that finished-product testing was not sufficient to assess the quality and sterility of parenteral products. Production and control procedures had to to be established to document that systems and procedures were under control. The term *validation* was used to describe the establishment of "documented evidence" that manufacturing processes of parenteral dosage forms consistently met defined quality characteristics. The initial validation concept focused on the sterilization process, aseptic operations and the process used to generate water in the preparation of the product. But it was soon recognized that other critical systems associated with the manufacturing process (e.g., heating, ventilation and air

277

conditioning and media like compressed air and nitrogen) had to be included in the concept as well.

In the 1980s, the validation concept was further developed and extended to encompass non-sterile dosage forms.

In 1978, the Food and Drug Administration (FDA) introduced validation into the revision of the clinical Good Manufacturing Practices (GMP). In May 1996, the FDA published proposed amendments in the *Federal Register* explicitly incorporating process validation including equipment qualification (sec. 211.220) and methods validation (sec. 211.222). In the highlights of the proposed rule, the FDA commented that "process validation is a *quality assurance* function that helps to ensure drug product quality by providing documented evidence that the manufacturing process consistently does what it purports to do."

In Europe, the validation concept arrived in the early 1980s. The requirement of validation was finally laid down in the Commission Directive "Guidelines of good manufacturing practice for medicinal products for human use" (91/356/EEC), published in 1991.

In Japan, the Ministry of Health and Welfare (Koseisho) has introduced validation as a legal requirement in the document "Validation Standard", effective April 1, 1996.

A summary of the relevant regulations for validation is given in Table 10.1.

Although validation has long been regarded as a regulatory must, a considerable amount of knowledge and experience has been gained through the years. Today, validation is well established as a valuable tool of an effective quality assurance system. There is no doubt, however, that validation has the potential to consume enormous resources, if not carried out with common sense and a pragmatic approach. Therefore, the strategy of implementation and integration of the validation concept must be well designed so as to prevent escalation of the resources consumed.

VALIDATION ACTIVITIES

Successful validation is a key part of the quality assurance (QA) system that ensures facilities, equipment and ancillary systems, procedures, processes and products are maintained in a state of

---

Table 10.1. Overview of legal requirements and guidelines for validation.

Regulations

- EU Commission Directive GMP 91/356/EEC (1991)
- USA 21 CFR Part 210/211 Current Good Manufacturing Practices
- Japan Ministry of Health Validation Standard (1996)

Guidelines and Guides

- EC Guide to GMP (1989)
- PIL/Document PR 1/99–1 "Recommendations on Validation Master Plan etc." (March 1999)
- PIC/S Document PS/W 4/97 Draft Recommendations on the Validation of Aseptic Processes (Feb. 1997)
- Guideline on General Principles of Process Validation (1987)
- Guideline on Sterile Products Produced by Aseptic Processing (1987)
- Guideline on Submitting Documentation for Sterilization Process Validation in Applications for Human and Veterinary Drug Products (1994)
- FDA Guides to Inspection of
  - Cleaning Validation (1993)
  - Pharmaceutical Quality Control Laboratories (1993)
  - Lyophilization of Parenterals (1993)
  - High-Purity Water Systems (1993)
  - Microbiological Pharmaceutical Quality Control Laboratories (1993)
  - Oral Solutions and Suspensions (1994)
  - Oral Solid Dosage Forms Pre/Postapproval Issues (1994)
  - Computerized Systems in Drug Processing (1983)
- FDA Validation Documentation Inspection Guide (1993) (this guide has not been oficially issued by FDA)
- PDA Technical Reports (e.g., No. 22 "Media Fills"; No. 25 "Blend Uniformity Analysis"; No. 29 "Points to Consider for Cleaning Validation")
- WHO Guidelines on the Validation of Manufacturing Process (1993)
- FIP Guidelines for Good Validation Practice (1980)

GMP compliance. Table 10.2 presents a summary of the GMP elements and the corresponding validation activities.

Validation Terminology

Validation terminology has become a matter of interpretation and argument, and there is often confusing discussion of the correct meaning of a term. It is therefore helpful to clearly describe and define what we understand by the different validation activities.

The term *validation* should be used to encompass all activities represented in Figure 10.1.

Table 10.2. Summary of GMP elements and validation activities.

GMP Elements	Validation Activities
Facilities, equipment, and ancillary systems	Qualification, maintenance, and calibration
Manufacturing processes	Process validation
	Validation of aseptic operations
Process control systems and documentation systems	Computer validation
Cleaning procedures	Cleaning validation
Analytical methods	Methods validation
Plant hygiene	Validation of sanitation procedures
	Ongoing microbiological monitoring of environment and utilities
Personnel	Personnel training
Documentation	SOPs for validation
	Documentation system for validation (protocols, reports, archiving of documents)
Process deviations	Application for deviations during validation
OOS—Failure investigation	Application for OOS during validation
Change control	Reevaluation/revalidation
Self-inspections	Ongoing monitoring of validated status
Vendor/supplier	Evaluation of suppliers
Contract manufacture and testing	Auditing and monitoring of contractors

Figure 10.1. Organigram of validation.

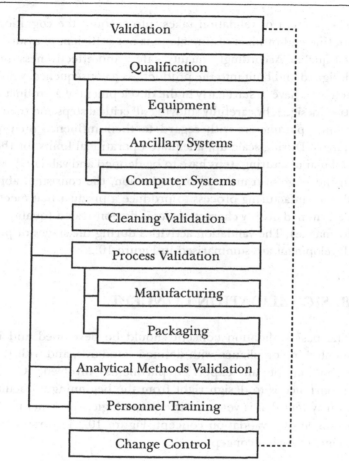

As a QA element, successful validation implies that processes and procedures, as well as ancillary systems, are well understood and under control.

On the other side, validation is not only a requirement of GMP but also itself a valuable tool for assuring that other QA elements— as far as they are relevant for validation—have been properly established. That means that the implementation of a validation concept is simultaneously a challenge of the QA system.

SCOPE OF VALIDATION

The concept of validation takes into account the cognition that routine end-product testing alone is not sufficient to assure product quality. Accordingly, quality, safety and effectiveness must be designed and built into the product. As a consequence, validation activities have to start early in the development of a product (quality of design), by carefully studying all critical steps and their operational parameters with regard to their influence on product quality. During scale-up, adequate operational limits for the critical manufacturing steps have to be defined and validated. Finally, in the stage of transfer into production, the consistent ability of the manufacturing process to produce a product that meets predetermined quality characteristics is demonstrated (quality of performance). The validation activities during the stages of product development are summarised in Figure 10.2.

BASIC VALIDATION CONCEPT

The basic validation concept should be developed and implemented in compliance with defined standards, and with the participation of all expert departments involved. Of major importance is to design right from the beginning a documentation system that is very consistent and logical in structure. As an example of a validation concept, Figure 10.3 illustrates the elements of such a concept.

STRATEGY FOR THE IMPLEMENTATION OF VALIDATION INTO A QA SYSTEM

The strategy for the integration of validation must be based on the company's quality policy and the commitment of senior management to support all validation activities as part of the company's quality policy. Based on that policy, the general concept of validation should be laid down in a corporate or company validation policy.

Figure 10.2. Stages of process validation.

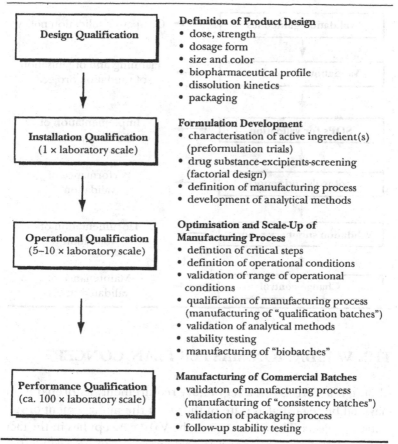

Design Qualification

Definition of Product Design
- dose, strength
- dosage form
- size and color
- biopharmaceutical profile
- dissolution kinetics
- packaging

Installation Qualification
(1 × laboratory scale)

Formulation Development
- characterisation of active ingredient(s) (preformulation trials)
- drug substance-excipients-screening (factorial design)
- definition of manufacturing process
- development of analytical methods

Operational Qualification
(5–10 × laboratory scale)

Optimisation and Scale-Up of Manufacturing Process
- defintion of critical steps
- definition of operational conditions
- validation of range of operational conditions
- qualification of manufacturing process (manufacturing of "qualification batches")
- validation of analytical methods
- stability testing
- manufacturing of "biobatches"

Performance Qualification
(ca. 100 × laboratory scale)

Manufacturing of Commercial Batches
- validaton of manufacturing process (manufacturing of "consistency batches")
- validation of packaging process
- follow-up stability testing

An essential part of the validation concept should be constituted by the establishment of a Validation Master Plan (VMP) for planning and execution of validation projects. A VMP is a document that summarises the validation programme based on the firm's validation philosophy. The establishment of the validation policy and a VMP provides all necessary elements for the integration of a validation concept in compliance with regulatory requirements (see Figure 10.4).

Figure 10.3. Validation concept.

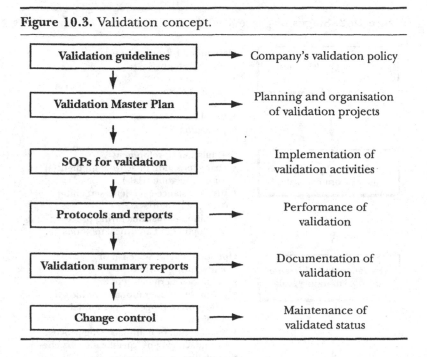

Validation guidelines ⟶	Company's validation policy
Validation Master Plan ⟶	Planning and organisation of validation projects
SOPs for validation ⟶	Implementation of validation activities
Protocols and reports ⟶	Performance of validation
Validation summary reports ⟶	Documentation of validation
Change control ⟶	Maintenance of validated status

THE VALIDATION MASTER PLAN CONCEPT

Although a VMP is not a regulatory requirement, its preparation may well be regarded as state of the art in the management of validation projects. The advantage of the VMP concept lies in the fact that it offers a systematic and logical approach to how validation activities can be planned, organised, monitored and documented. Even if some isolated validation experiments have already been performed, the VMP concept can be introduced to structure the validation programme company wide.

The establishment of a VMP urges a company to clearly describe and plan all necessary validation subjects, and to define responsibilities as well as the resources necessary for the implementation and execution of a validation programme.

As guidance documents for the preparation of a VMP, the PIC/S Document PRI/99–1 and the unofficial Validation Documentation Inspection Guide of the FDA can be consulted.

Figure 10.4. Quality policy pyramid.

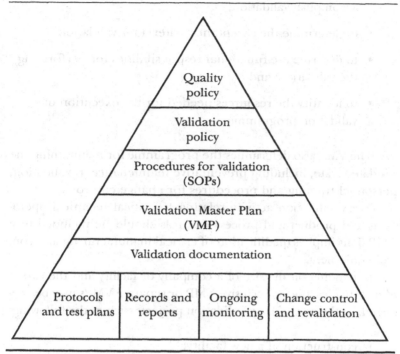

Purpose and Scope of a VMP

According to the PIC/S document PR 1/99-1 (March 1999)[1], a VMP should be a summary document and should therefore be brief. It should facilitate an overview of the entire validation operation, its organisational structure, its content and planning and its core should be on inventory of the items to be validated and the planning schedule.

Accordingly, the VMP serves as a project management tool to assure that the validation project is planned and carried out in a structured and controled way.

The purpose of a VMP includes the following points:

- to identify which systems, equipment, processes, and procedures are subject to validation;

- to define the nature and extent of testing to be done on each item;

---

[1]Recommendations on Validation Master Plan etc.

- to outline the procedures and tests to be followed to accomplish validation;

- to determine the acceptance criteria for validation;

- to describe the functional responsibilities for performing the validation; and

- to identify the resources needed for the execution of validation programme.

The VMP also determines the programme for maintaining the validated state, including preventative maintenance, revalidation, personnel training and procedures for change control.

Every validation activity relating to critical technical operations and product and process controls should be included in a VMP. This implies qualification of critical manufacturing and control equipment.

Depending on the size of a company or facility and the extent of its activities, one all-inclusive VMP or separate VMPs may be preferred. In the case of large validation projects such as the following:

- construction of a new facility;

- reconstruction of existing facilities;

- validation of cleaning procedures;

- validation of computer systems;

- qualification in the QC laboratory; and

- scale-up of new products and transfer into production.

The best approach is often to create a separate VMP. In any case, the VMP should be based on the firm's overall philosophies and intentions and should be approved and supported by senior management.

It should be emphasized that a well-designed VMP may also serve as a valuable document during regulatory inspections, not only to provide information and documentation about the validation project, but also to demonstrate the company's commitment to complete validation work, in the case that not all activities have been completed at the time of an inspection.

OUTLINE OF A VMP FOR A PHARMACEUTICAL PRODUCTION PLANT

This section presents the structure and content of a VMP which complies with the guidance documents mentioned above and has been proven in practical application (see Table 10.3).

The content of the different chapters of the VMP are briefly described and commented on.

Preamble

The preamble should briefly describe the overall validation policy of the company, and include a statement by senior management to support the validation programme as described in the VMP and to allocate the necessary resources for the execution of the programme.

Table 10.3. Structure of a VMP.

---

1. Preamble
2. Introduction
3. Description of the validation programme
4. Responsibilities
5. The Validation project plan
6. Validation organisation
7. Handling of deviations during validation
8. Programmes for maintenance of validated status
9. Appendices

Introduction

In the introductory chapter the objectives of the VMP, the applicable regulatory standards, and provisions for the revision of the VMP should be included.

Description of the Validation Programme

This chapter may contain the following information:

a. General principles of validation, defining company-specific strategies and procedures (as described in the validation guideline) like teamwork, risk analysis, procedures for new vs. existing equipment, contracting out of validation work and so on.

b. Description and scope of validation activities, including facilities, support systems, utilities and equipment, products and procedures and computer systems.

c. General acceptance criteria for all validation activities.

d. Description of the Standard Operating Procedures (SOPs) for validation. A list of the relevant SOPs would be enclosed in the appendix of the VMP (see list of SOPs in Table 10.4).

Responsibilities

Usually, the planning and execution of validation projects involves a teamwork between different expert departments. In this chapter, the responsibilities of the Validation Team and the participating expert departments are described considering legal requirements. Those experts include members from Production, Quality Control (QC)/QA, Engineering, and Development. It is important that the team be involved at the time the design commences and have interactive discussions throughout the whole validation project, in order to streamline the validation activities and avoid duplication of work.

Table 10.4. List of SOPs for validation.

Validation policy of the company or corporation

Procedure for the preparation of protocols and reports

Procedure for qualification of equipment and support systems

Procedure for process validation

Procedure for cleaning validation

Procedure for computer validation

Procedures for maintenance and calibration

Procedure for personnel training

Change control procedures

Procedure for deviations in production

Procedure for validation of analytical methods

Procedures for qualification of analytical instruments

Procedures for calibration of instruments, and apparatuses

Procedure for sampling during cleaning validation

Procedure for microbiological monitoring of production areas

Procedures for monitoring of water systems

Procedures for control of aseptic operating conditions for technical equipment and personnel (media fills)

Procedure for monitoring of sanitation in sterile areas

Procedure for bacterial challenge test for the validation of sterile filtration

Procedure for validation of steam sterilization in autoclaves by aid of bioindicators

Procedure for validation of depyrogenation in a dry-heat sterilization tunnel

Procedure for validation of steam sterilization of solution preparation vessels

Procedure for out-of-specification results

This is *not* an all-inclusive list. Other SOPs may also apply.

The Validation Project Plan

The basic concept of this chapter is to identify, in the form of a risk assessment, all relevant validation activities related to the project. Therefore, the objectives of this chapter are the following:

- To provide a detailed description of the validation project. The project description should include facilities, equipment, and support systems, as well as the hygiene zone concept, including flow of materials and personnel, products and procedures handled in the facility. The desription should be illustrated by layouts, schematic drawings, flow diagrams and so on.

- To establish a list of all items subject to validation based on the risk analysis, including the respective activities (e.g., for equipment and systems the Design Qualification (DQ), Installation Qualification (IQ) and Operational Qualification (OQ)).

- To prepare the validation project plan. This project plan is intended as a timeline for all validation items to ensure that the validation programme is executed as planned. This project plan is the true project managment instrument, and should therefore be periodically revised and updated.

The validation project plan could be established using a suitable software programme to facilitate follow-up and updating of the project status. If appropriate, the validation project plan could also be employed for planning and scheduling of the resources necessary for the validation project.

Validation Organisation

The organisation of the validation project includes the definition of the procedure for the documentation of validation, and the forms to be used for protocols, test plans, reports and so on.

A system for the documentation of validation must be established. Usually after approval of the documents, files are prepared for each validation subject and stored in an archive from which they are easily available.

Handling of Deviations during Validation

In this chapter, provisions should be described in the event of deviations. Such deviations may encompass the following cases:

Deviations

Deviations may be defined as unplanned deviations from protocols or test plans due to unforeseen difficulties in the instruction, or malfunction of equipment or operator failures. The handling and documentation of any deviation should follow the established procedure (SOP).

Out-Of-Specification Results (OOS)

OOS may occur when the result of an experiment does not comply with the acceptance criteria as defined in the protocol. A differentiated procedure should be defined for failure during qualification, process validation and so on. The handling and documentation of OOS results should follow the established SOP.

Changes

Changes are planned alterations to protocols or test plans which may occur when a planned experiment is not feasible. Changes are usually approved before realization.

It is important to describe the procedure for the handling of the above mentioned deviations with regard to approval (or disapproval) and what consecutive actions must be taken. Usually, the validation team is the responsible committee for the handling of deviations. All deviations have to be documented, cited, and commented on in the validation reports.

Programme for Maintenance of Validated Status

An essential element of the VMP is a programme to assure that the validated status of facilities and equipment ancillary systems, as well as procedures and processes, is maintained in future. For that purpose, the VMP defines activities embracing environmental monitoring, preventive maintenance and calibration, and a system of change control and revalidation. The personnel training programme supplements this programme.

Environmental Monitoring

Here the procedures applicable for the monitoring of nonviable and viable particulates in the GMP areas should be inserted. Refer to SOPs.

Monitoring of Water Systems

Insert here the procedures applicable for the physical-chemical and microbiological monitoring of the PW and WFI sytems. Refer to SOPs.

Preventive Maintenance and Calibration

A description of the preventive maintenance programme which is in place to support the ongoing qualification requirements and assure the validated status of machinery and equipment should included here. Refer to SOPs.

Re-Validation

Describe the established procedures. Refer to SOPs.

Change Control Procedure

Describe the established procedures. Refer to SOPs.

Personnel Training

In addition to the regular training programme, a specific programme should be executed to familiarize all staff involved in validation with all relevant aspects of validation and the specific requirements of the validation project.

Appendices to the VMP

Usually the following documents are enclosed in a VMP:

- Project plan, including all updates
- Site plan
- List of applicable standard operation procedures
- General plans/layouts of facilities
- Plans/layouts showing flow of materials and personnel
- Scheme of HVAC system
- List of products
- Flow or block diagrams of processes
- List of manufacturing equipment
- List of packaging equipment
- Forms to be used for validation (protocols, reports)
- Revisions/amendments to the VMP

In cases where separate VMP are to be prepared for the validation of cleaning procedures and for computer validation, such documents may also be enclosed in the appendix.

BENEFITS OF VALIDATION

In the first place, validation is a regulatory requirement and a prerequisite to meeting the expectations of regulatory inspections. But in recent years, criticism arose in the pharmaceutical industry regarding the extent of validation activities and the documentation generated during validation (Sharp 1995; Akers 1993, 1995; Anisfeld 1994). In many cases, very little consideration has been addressed to the benefits for the patient and the pharmaceutical manufacturer.

There is no doubt that validation has the potential to consume enormous resources, if not carried out with common sense and a pragmatic approach. On the other hand, a reasonable amount of knowledge and experience has been gained in this area through the years. A validation concept designed with common sense and a company-tailored approach should be able to prevent escalation of resources and may well serve as a valuable tool to ensure the quality of pharmaceutical products.

The benefits of validation as a QA tool may be described as follows:

- Better knowledge of the products to assure quality and safety
  - specifications for active and critical inactive ingredients
  - stability of products
  - certified vendors for materials
- Better knowledge of the processes to avoid product failures
  - challenged operating conditions with regard to worst-case conditions
  - reproducibility of critical processes such as sterilization cycles
  - controlled environmental conditions
- Suitable equipment and support systems to increase productivity
  - qualified equipment
  - implementation of preventive maintenance
  - effective cleaning procedures
  - approved computer-controlled systems
- Trained personnel to operate and maintain validated processes and equipment

- Development of personnel's skills and experience

- Improvement of operator's habits, motivation and work organisation

- Implementation of teamwork for better mutual understanding

- Compliance with regulatory requirements for GMP and QA

  - Compliance with GMPs

  - Assurance of product licences

  - Avoidance of delays in new drug applications

  - Controlled procedures for changes

- Systematic documentation

  - Starting point for improvements

  - Statistical trend analysis for process evaluation

  - regular review and evaluation to ensure quality standards

Economically, if a company can reduce destruction or rework (or even recalls) of failed batches, the costs of validation may easily be recouped.

Therefore, a well-designed validation programme based on both scientific principles and common sense is a valuable tool of a quality assurance system and a source of progress.

REFERENCES

Akers, J. 1993. Simplifying and improving process validation. *Journal of Parenteral Sciences and Technology* 47 (6): 281–184.

———. 1995. Validation—Does it do what it purports to do? *Imagin-Action* 6: 111–118, Italy: I.M.A. Idustria Macchine Automaticche S.p.A.

Anisfeld, M. H. 1994. Validation—How much can the world afford? Are we getting value for money? *J. Pharm. Sci. Tech.* 48 (1): 45–48.

Demmer, F. "Validation Master Plan" in Concept Seminar Validation Master Plan, 23.02.1999, Mannheim.

Sharp, J. 1995. Validation—How much is required? *J. Pharm. Sci. Tech.* 49 (3): 111–118.

11

How to Build a Supplier Qualification System

Wolfgang Schumacher
ASTA Medica AG
Frankfurt, Germany

QUALIFICATION OF SUPPLIERS

These days, the business press is filled with new management theories for revitalizing worldwide business. The common thread running through all of them is the need for companies to rethink the way they operate. They must become more efficient and productive to remain competitive in the global marketplace.

The increasing pressure on the pharmaceutical industry to bring new products to market quickly and for reasonable cost has resulted in many efforts to reduce expenditures. The significant changes in this branch—with many mergers, acquisition and strategic alliances—influenced in particular the partnership with suppliers[1], contract manufacturers and service suppliers. Fair supplier evaluation is a success factor for good long-term business relations. Is it possible for a company to be better than its suppliers? The question of how to best measure and evaluate supplier performance is one of the key problems for the purchasing, quality and production departments.

This chapter aims to discuss the possibilities and facts to consider when introducing a supplier qualification system in a pharmaceutical company.

General Aspects

The aim of systems for supplier evaluation is the transfer of the various factors influencing the quality into easy-to-measure parameters. But, before thinking about such a project, basic company strategies should be defined, including the following:

- the general purchasing strategy;

- strategy on single or multiple sourcing;

- the supplier approval process; and

- the importance of specifications.

The Purchasing Strategy

A purchasing strategy serves the overall business objectives and business strategy. It may encompass the following:

- Building partnerships with suppliers to get the best deal for the firm, in the short and long terms. Through such partnerships the firm can better secure its important supplies with the right quality, the right delivery terms, and at the lowest cost.

- Using the know-how of the suppliers to contribute to business planning, helping to predict changes in markets and technologies.

---

[1]The terms "vendor" and "supplier" are used in this chapter in the same sense.

It certainly means that the following points should be considered:

- Supply industries' capacities in comparison with customer Requirements.

- The current and future competition for supplies.

- Supplier order book position relative to own position.

- The existence of any monopolizing power in the market.

- The size of the supply markets compared with one's own market share.

- Suppliers' ability to meet present and future demands.

With the overall business strategy, it is necessary to develop a strategic plan for purchasing where the supplier qualification plays an important role. As well as the specific objectives, the purchasing operation will be conditioned by basic business objectives, such as the following:

- Minimize the cost of supplies.

- Improve quality and reliability.

- Ensure that supplies are always available when and where needed.

- Minimize investment in inventories.

- Evaluate the market for new and improved products and services.

Although common to nearly all branches, the degree of importance attached to these basic objectives will vary according to the individual circumstances. For the pharmaceutical industry the quality is of significant importance.

The most important things must be considered first. Sometimes it seems unclear what the important things are because of different viewpoints. In many companies prioritizing is a major task in itself, so the information systems should be tested to see if they can identify answers to the following questions:

- What was the total value of bought-in supplies (including contract manufacture) last year?

- How many suppliers does the firm have?

- What were the average expenses last year?

- Who were the top ten suppliers by value?

- What were the ten most expensive items bought last year?

- What are the quality requirements for the most important buys?

Single or Multiple Sourcing?

Where a particular supplier has a monopoly—for historical or technical reasons—there is no big choice. But in most cases there are alternative sources. It is useful to have a general policy position on single or multiple sourcing but, as everything hinges on the prevailing conditions in a market, each option must be evaluated. Unfortunately, there are no general recommendations possible.

Aggregating the business with one supplier should give the customer a better price, reduce necessary additional costs and improve the service received. On the other hand, the competition resulting from dividing the business among two or three suppliers could in total result in a better deal. A well-motivated, single-source supplier may be a more secure one in difficult trading conditions. Alternatively, a second source will give greater security in the event of an accident or other unforeseen events at the main supplier. Strong arguments can often be made on both sides, but the trend among successful firms is toward reducing rather than expanding the supplier base, i.e., in the pharmaceutical business one alternative supplier for all critical raw materials.

Product Specifications

The specifications of the product must be set to adequate levels. Therefore, all persons involved must help establish and review such data. Particular reference must be made to the legal requirements and the registration file deposited with the authorities. An intensive and detailed discussion will help to relax certain less-critical specification parameters, if this is in compliance with the registration file, for example, for suppliers of packaging material. Tighter specifications will result in higher yields, a more consistent product quality and reduced equipment downtimes.

Changes to those specifications must be documented with the supporting data, and the customer must approve all major changes before they are realized by the supplier.

Reduced Testing

The reduction of the testing is one of the most supporting reasons to set up a supplier qualification program. Under certain circumstances the health authorities agree to take over analytical results of the supplier. As the cost of quality operations has become a very important factor in pharmaceutical manufacturing, the elimination of some testing is considered to be positive, but even more important is the fact that materials will become available to production more quickly. This allows reductions in inventory and is beneficial as well if items are urgently needed for unforeseen production or in events.

THE QUALIFICATION CONCEPT

The Supplier Evaluation Team

It is mandatory to set up the responsibilities for the supplier qualification system from the beginning. Several functions need to be involved in establishing the vendor evaluation team. The initial task of the team should be to define the objectives and potential benefits and to describe a process which can be used as a basis for the discussion with suppliers.

Typically, the team will include representatives from manufacturing, package engineering, purchasing, quality assurance
(QA) and quality control (QC) with support from other disciplines
(for example, research and development), if necessary.

The *coordinator* should be the specialist for the company, the
market, and the pricing situation, preferably a member of the purchasing department.

The *product specialist* (manufacturing, package engineering)
knows the item to be purchased, the process and the regulatory
situation.

QC has all information on specifications and test results.

QA has evaluated the supplier during the audit and knows best
how the supplier will meet the requirements in the future.

Supplier Performance Criteria

The most important step of the future qualification is to develop
a clear-cut concept for the requirements of the company. The formalized scoring system (see the "Supplier classification and rating systems" section) has to be implemented for all possible kinds
of suppliers. All non-conformities are to be evaluated by the different departments.

The Selection Process for a New Supplier (see Figure 11.1)

If it has been confirmed via the primary evaluation of purchasing that a supplier has established a controlled process, the
quality of the product is checked. Usually, pharmaceutical manufacturing operations use "process qualification runs", i.e., pilot
runs, under the same conditions and sometimes even with the
same equipment as will be used in production. This concept has
been proved to be the best one when a new raw material is tested
in the production equipment. After comparing the data and positive results obtained there, audit the suppliers' facilities to get a
complete picture of the company. That site visit will be not necessary if there is comprehensive information available (see the
"Questionnaires/Checklists" and "Company Information via FOI"
sections). As a result of the audit (see the "Audits of Suppliers"

Figure 11.1. Management of suppliers.

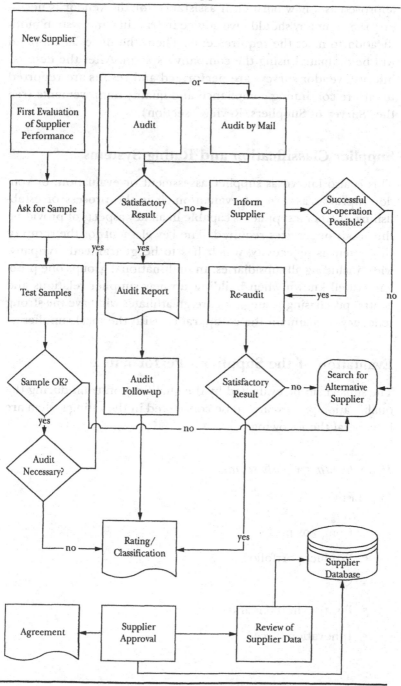

section) and the defined follow-up activities, the supplier will be approved as a new partner. If assistance for the vendor is necessary the company should give advice in training and system modification to meet the requirements. Then, the material supplied will be evaluated using the company's system. After the defined interval, vendor surveys are performed and results are reported to ensure continuing compliance and quality improvements (see the "Survey of Suppliers/Review" section).

Supplier Classification and Rating Systems

This is also known as supplier assessment or evaluation, or vendor appraisal/assessment/evaluation. It is the process of establishing whether a supplier is capable in all key aspects of providing the goods or services required. The key element for the setup of the system is objectivity, which has to be guaranteed company-wide, including all subsidiaries. In multinational groups one problem is well known when building up new supplier relations and central purchasing strategies: Foreign affiliates will have the strong tendency to maintain the co-operation with the local supplier.

Evaluation of the Supplier's Performance

The subjects to be evaluated by the qualification team during the qualification process and to be considered in the rating system are in general the following:

Prior to Start Co-operation

Purchasing

- Company profile
- Product portfolio
- Sites
- Position in the market
- Innovation force

- Stability of the management
- Qualification/education of the staff
- Customer orientation
- Flexibility

During Co-operation

- Purchasing
- On-time delivery
- Completeness of shipment
- Service
- Long-term reliability
- Partnership
- ISO certification

Logistics

- Proper condition of the consignment
- Completeness of shipping documents
- Outer packaging
- Labeling of the different single packs

Quality Control

- Certificate of analysis
- Conformity to specification
- Inner packaging

Production

- Suitability for the process
- Output, yield

- Packaging

- Exact quantity

- Quality Assurance

- Quality assurance agreement

- Result of the on-site audit

- Complaints

Example 1

The following low-cost evaluation system successfully used in the pharmaceutical industry has been published in a small booklet by the German Drug Manufacturers Association (BAH 1996).

Using simple standard evaluation sheets, incoming goods are assessed in the plant by the departments. The five main criteria—quality, co-operation/service, reliability, cost/price and experience with the supplier—are checked for each delivery.

The results of the individual checks of the sub-criteria (e.g., complaints under quality) are filled in by the operators; then the main criteria are calculated from these results by QC. Zero to 10 points are to be attributed to the main criteria (10 points = 100% satisfaction). The final evaluation is performed by Purchasing and QC.

The classification of the suppliers according to this system is shown in Table 11.1.

If certain criteria with influence on the product quality have been assessed to be 0 (zero), the supplier is downgraded to C (not approved).

Example 2

A much more complicated evaluation system has been developed by the electronics industry on the basis of the business model of the European Foundation of Quality Management (EFQM), Brussels (Huck 1996).

The quality of the supplier is evaluated by checking a catalogue of elements in different modules. Therefore, the flexible system must be adopted to the individual needs by weighting those

Table 11.1. Classification of suppliers.

Overall rating	Evaluation	Class
90 to 100	Completely satisfactory (approval)	A
80 to 90	Mainly satisfactory (limited approval)	AB
60 to 80	Partially satisfactory (conditional approval)	B
< 60	Not satisfactory (no approval)	C

elements which are allowing some flexibility. Some pharmaceutical companies are using similar custom-build rating systems, such as that shown in Table 11.2.

Example 3

The evaluation of the suppliers' performance with respect to the main criteria—quality, quantity, and delivery date—is successfully used in the health and medical device industry as well.

Quality. The results of the income inspection and QC tests of the product are the basis for the evaluation. If the product does not meet the agreed specifications, the different classes of non-conformance are applied; they are weighted differently. The total number of shipments (or products) received over a defined period (e.g., 2 years) is used for the calculations.

Quantity. The vendors' reliability to supply the quantities ordered and confirmed is evaluated. The performance regarding quantity is classified to be the deviation of the quantity delivered from the quantity ordered.

Delivery Date. The reasons mentioned above under the term "quantity" are also applicable for the evaluation of the delivery time. Depending on the logistic consequence, the deviation (in weeks) of the real shipment from that confirmed by the supplier is considered.

The overall rating of the supplier in the classes A, B and C is done in principle, as mentioned in Example 1.

Table 11.2. Categories and general criteria to build a modular supplier management system.

Company	Technology	Process	Products	Shipping
Profile	Product strategy	Documentation	Quality	Logistics
Management	Production	Planning	Reliability	Transport system
Quality system	Development	Environment friendliness	Complaint management	Cost management
Co-operation, Service				

Example 4

The vendor approval status indicates that the supplier has been classified according to the established system. Another possibility to define the different classes of suppliers is the rating system used in the pharmaceutical industry, shown in Table 11.3. The mathematical approach of the systems mentioned so far is replaced by a pragmatic, paper-based approach which does not need support of a sophisticated computerized system or database.

Survey of Suppliers/Review

After selecting a suitable supplier, it is important to maintain the commitment to the qualification project, i.e., both parties should routinely review the progress and the existing problems as well. In the quality-oriented pharmaceutical manufacturing world, many companies require more stringent QCs and have specifically developed very detailed expectations as to how their vendors will set up their systems. Most of these require that a company's vendors also meet the same standards. It is also essential that any vendor be closely surveyed, but most medium to small companies do not have the staff nor the systems in place to evaluate their vendors or to do the required surveys to certify them. Both partners should be aware that the handling of bad news, which can very easily come up in the review period, is not an easy task (see Figure 11.1).

Table 11.3. A typical rating system used in the pharmaceutical industry.

Class	Criteria
A	Audited, 3 consecutive consignments without non-compliance, on-time delivery;
	process and manufacturing company is known (in case of broker/trader); and
	commitment to inform customer in case of process changes.
B	Same criteria as A class, but without audit; or
	former A-class supplier with the need to optimize certain criteria.
C	Qualification program required due to low performance regarding important criteria; and
	necessity of close monitoring.
D	New supplier;
	analytical quality and pilot batch approved;
	process known; and
	change control commitment signed.

ERP Systems

Many Enterprise Resource Planning (ERP) systems offer the possibility to generate their own vendor appraisal systems or to provide interfaces to separate systems which ensure a transparent and fair rating of suppliers and contractors. SAP R/2 (from Release 5.0) and R/3 offer a complete set of criteria for the evaluation of suppliers in the QSS module (R/2) and the QM module (R/3).

These systems are based on the systematic creation of a qualified-suppliers database built up over a period of time as procurements are undertaken. A supplier's status (e.g., accredited, qualified, pending, rejected or suspended) is linked to a particular item or product group. The details within each product group include ratings of the supplier against the criteria mentioned previously (financial, technical, quality, commercial and environmental). These main criteria can be further subdivided into user-definable criteria, for example, credit standing, audit index and so on.

The score for each supplier should be calculated automatically, but there should be the possibility to override that calculation if a reason for the change in rating is given.

Audits of Suppliers

Vendor audits are not required in every case by the Health Authorities, but they are strongly recommended. With regard to the supplier-evaluation process, the results of audits cannot be the only criteria to measure the performance of the vendor.

As audits are very time-consuming, many consulting organizations offer to carry out such evaluation visits with specific scientific know-how in the business field of the audited. For the responsible pharmaceutical entrepreneur, using a consultant is very often the only way to perform complicated audits, for example, for the computerized systems to be purchased. The "shared audits" or "joint audits," in which some companies bring their experience together are beneficial ways for both the supplier and the customers: The supplier can save considerable time by hosting some customers together, and the auditors by exchanging their knowledge. A problem might be the confidentiality of new technologies which are sometimes discussed on the occasion of audits.

Some groups of pharmaceutical companies have existed in Germany for many years, like the Bonn German Drug manufacturers Association BAH (Bundesfachverband der Arzneimittelhersteller), the HK Nord (Herstellungs-und Kontrolleiterkreis Norddeutschland) and the HK Süd (Herstellungs-und Kontrolleiterkreis Süddeutschland).

Company Information via FOI

If there is not enough capacity and time to perform a full vendor audit, the information available via the Freedom of Information Office (FOI)[2] of the Food and Drug Administration (FDA) can be very useful. The Establishment Inspection Report (EIR), which is

---

[2]5600 Fishers Lane, Rockville MD 20857

usually a detailed description of the activities of the site and the 483s (if any) of the last inspection, give a complete "quality picture" of the supplier or contract manufacturer. Both documents can be requested in writing for a small fee covering the search, copy, and mail by the FOI office. (Allow approximately 3 weeks for documents to be mailed; older reports can take up to several months.) For primary information on a future contract partner the internet page of FOI within the fda.gov server can provide information on warning letters; if the contract partner has obtained such a warning letter recently, the reason should be evaluated carefully. A written statement of the company explaining the follow-up activities is mandatory to protect the own company.

Questionnaires/Checklists

The use of questionnaires and checklists is popular in the pharmaceutical industry. These serve for the primary evaluation of the quality and business basics of suppliers and contract manufacturers as well as for use during the audit.

The questionnaires are submitted to the new vendor to check the presence of a quality management system (QMS) (audit by mail). Some of the pharmaceutical manufacturer associations have developed such standard supplier quality assessment questionnaires to be completed by chemical vendors. Most of these documents require the company to provide all information requested; therefore, quite often the information given is not detailed enough to enable an assessment as a substitute for an on-site visit. In many cases vendors are reluctant to disclose such information at all in writing. Some self-audit questionnaires provide only short multiple-choice answers for every question (yes/no/NA); limited benefit can be drawn from such information. The only benefit of the mail audit is to serve as a good preparation for the on-site audit.

Checklists to be used during the audit are helpful for auditors with less experience. A complete set of such checklists with strong emphasis on the Food and Drug Administration (FDA) regulations and International Standardization Organization (ISO) 9000 series is available, for example, by Interpharm Press (1995). Checklists also provide ratings to measure the different criteria during the audit. They are rating performance, for example, from 0 (worst) to 5 (best). As the individual rating is strongly dependent from the auditors' experience it is mandatory to explain the weak points and positive impressions evaluated during the audit in a detailed report accompanying the checklists. This report also gives the recommendation (or not) to co-operate with the audited organization in the future.

The Supplier Agreement

The major definitions of the future co-operation with the supplier should be clearly outlined in a written agreement. For the outsourcing process of certain activities to a contract manufacturer the agreement is mandatory (except United States). As many financial details have to be defined with the contract partner as well, it is beneficial to separate the commercial contract from the technical/QA agreement.

Such technical agreement should contain:

- Description of the general contractual situation

- Information on the process

- Change control information procedure

- In-process controls

- Final controls of the supplier

- Logistics

- Complaint management

- Handling of sub-contractors

- Secrecy

The *annex* to the contract should contain information which can easily be updated:

- Specification of the product

- Acceptable Quality Level (AQL)

- Documentation required
- List of responsible persons and contact persons
- Manufacturing license
- ISO or equivalent certificates

The Role of ISO 9000

The systematic evaluation of suppliers is a must for all firms with ISO certification. There is also a set of formal quality tools which may be useful as well. A certificate from an accredited certification body shows that the supplier's system complies with the relevant part of ISO 9000. The certificate shows that the supplier has established and is maintaining a documented quality system as a means of ensuring that its products and services conform to specified requirements. It is no guarantee that the products are meeting in every case *all* specifications, but it should be most likely. If the potential supplier does not have such a certificate, someone familiar with assessing systems and with the technology could be contacted to define the potential supplier's capability.

In this context it is important to remember that the pharmaceutical entrepreneur always carries the responsibility for the product brought by him to market. A certified supplier will have systems for reviewing the requirements of the contract, but the pharmaceutical manufacturer must disclose the standards that have to be met.

Requirements of the Health Authorities

The Regulatory Bodies have established requirements for the manufacturers (CFR Title 21, §211.84), of the finished dosage form and their suppliers of Active Ingredients, Excipients, and other components. Some of those guides, including The Code of Federal Regulations (CFR) and the FDA Guides to inspection of. . . , the World Health Organization (WHO) (1996) and the European Economic Community (EEC) GMP Guidelines, the International Conference on Harmonization (ICH) (1999) and the International Pharmaceutical Excipients Council (IPEC) Guide

(1997) define the requirements and expectations on the basis of the components purchased and the services used. In the pharmaceutical and medical-device industry, the need to comply with the various regulations is mandatory as regular inspections of the Health Authorities are performed on site.

The direct relation and partnership between the pharmaceutical entrepreneur and the supplier are quite clearly defined for the contract manufacturer, for example, in the German Pharm-BetrV (§12). The EEC GMP Guide (EEC Guide to GMP 5.25, 5.40) requires that suppliers have been approved for starting materials and packaging items. Specifications and details of production and control are to be discussed with the supplier (EEC Guide to GMP 5.26). The WHO GMP regulations (1992) refer to "supplier audits", "supplier approval", and "supplier evaluation" (9.8, 9.9). Release on the basis of supplier certificates is deemed to be acceptable only if the results are validated periodically and "on-site audits" have been carried out.

Documentation

The system for the supplier evaluation should be defined in several Standard Operation Procedures (SOPs) and in the Quality Manual, if any, as well. If certification according to ISO 9000 or the upcoming QS 9000 series is planned, it will be part of the certification process and the respective audit. The basis in the firm has to be defined in the Quality Manual, including a commitment by management to take care of the important process of supplier evaluation to guarantee best drug product quality for the patient.

The Purchasing Department is authorized to issue, change, or cancel purchase orders and to maintain order information; significant orders that affect the product quality are checked for adequacy before release. Therefore, purchasing is responsible for—in particular—the following documents to be on hand:

- Records of supplier evaluations.
- Records/lists of approved suppliers.

- The SOP "Evaluation of Suppliers/Vendors" will define the detailed criteria used by the firm and the responsibilities for the contribution to the system (e.g., by QC, production, and purchasing departments). The process of collection, entering and review of material data is described there, and the different ratings (e.g., class A to D) and actions given.

- The SOP "Release of New Suppliers" shall define the process and the responsibilities to obtain acceptance for a new vendor. All departments involved will take part on the formal release process. It is recommended to define the individual tasks using matrices.

- The SOP "Audit of Suppliers and Contract Manufacturers" will clearly define the planning, performance, and follow-up activities for the audits to be carried out. It is recommended to limit the size and to define the responsibility of the lead auditor to set up the audit team.

- The audit program on a yearly basis will cover all partners to be audited. This program should be harmonized between the different locations in multi-site companies to avoid multiple visits of the same supplier.

- The "Non-Conforming Material Reject Report" or a similar document can be used to notify the purchasing department and concerned personnel of non-conforming materials and what action has to be taken. This report is applicable to all non-conforming materials. Purchasing will be responsible for all negotiations regarding financial costs and return methods (shipping) with regard to such material. QC/QA will be responsible for obtaining the prompt Corrective Action Report from suppliers.

Summary

QA of suppliers is necessary to get confidence to receive the goods wanted, performing as specified, at the right time. Certification results in a high level of reliance on the supplier. Vendor certification provides a strong basis for the application of reduced testing. Reduced incoming inspection, reduced inventories and higher output are the most beneficial effects. Since it is a partnership between the pharmaceutical company and the supplier, both parties should provide feedback that long-term co-operation is successful and confidence will be maintained. The introduction of sophisticated computerized supplier-rating systems should be checked carefully because they need considerable personnel and financial efforts by all departments involved. The outcome of a small and easy-to-handle paper-based evaluation system may contribute to the effects wanted in the same way. For contract manufacturers the application of the rating system is more complicated and usually does not bring real benefit.

REFERENCES

CFR Title 21, §211.84.

EEC Guide to GMP, 5.25 and 5.40.

EEC Guide to GMP, 5.26.

German "Betriebsverordnung für pharmazeutische Unternehmer", §12.

GMP/ISO 9000 Quality Audit Manual for Healthcare Manufactuers and their Suppliers, ISBN 0-935184-65-1, Interpharm Press, 1995.

Good Manufacturing Practice for Active Pharmaceutical Ingredients, Draft 3, ICH, February 1999.

Huck, W., Köpke, G., Faire Lieferantenbewertung, QZ 41 (1996) 10, 1171.

The International Pharmaceutical Excipients Council, *Good Manufacturing Practices for Bulk Pharmaceutical Excipients*, 1997.

Standardverfahrensanweisungen (SOPs) der fiktiven Firma "Muster" für die Arzneimittelherstellung, Bonn, 1996 (new edition, 1999, in press) Editor: BAH (Bundesfachverband der Arzneimittelhersteller).

World Health Organization (WHO) *Good Manufacturing Practices for Pharmaceutical Products*, 1992

WHO, *Good Manufacturing Practices for Pharmaceutical Products, Supplementary guidelines for the Manufacture of Pharmaceutical Excipients*, 1996.

12

Hazard Analysis and Critical Control Points in a Pharmaceutical QA System

Michael Jahnke

Pharma Hameln GmbH
Hameln, Germany

In order to guarantee the manufacture of consistently high-quality products for human use, it is absolutely essential that flawless hygiene conditions are maintained through the strict observance of hygiene rules. The purpose of such rules is to ensure that measures are taken to protect products from any type of contamination during the manufacturing process. The user of the product is thereby shielded from risks to health through chemical or biological impurities, the product is protected from any bioburden and the acceptance of the preparation is increased.

With the growing understanding of the impact of process conditions on the quality of resulting products, process controls have become an increasingly important part of the quality profile traditionally defined by post-process product testing. Process controls are today a standard Good Manufacturing Practice (GMP) requirement. However, before suitable process controls can be established, the manufacturing process has to be analysed in terms of its critical steps. Such a hazard analysis and the monitoring of Critical Control Points (CCPs) will lead to the establishment of a preventive monitoring system, and constitute an effective concept for quality assurance (QA) of hygiene and all other parameters influencing product quality.

As a multidisciplinary and structured analytical tool, Hazard Analysis and Critical Control Points (HACCP) is used to introduce and maintain a cost-effective, ongoing safety program in the pharmaceutical industry. HACCP involves the systematic assessment of all process steps involved in a manufacturing operation and the identification of those steps that are critical to the safety of the product or to the reproducibility of the product quality.

HACCP is applicable to the identification of microbiological, chemical and physical hazards affecting product safety, and can be used to establish control measures focused on product safety, in-process control and reproducibility of product quality. With respect to pharmaceutical safety and QA, the need for and extent of qualification and validation measures of equipment used in the pharmaceutical process are identified. Therefore, risk analysis in the pharmaceutical industry combines potential microbiological, physical and chemical hazards of a specific process/product combination with the the concept of Failure Mode and Effect Analysis (FMEA). FMEA is an engineering system which focuses on all components of a process and asks what might potentially go wrong within the manufacture process.

The HACCP approach allows companies to concentrate resources on those steps that critically affect product safety. As the result of such an analysis CCPs will be identified, for which limits, monitoring procedures such as in-process control measures and corrrective actions have to be established.

One of the many advantages of the HACCP concept is that it moves away from a philosophy of control based on end-product testing (i.e., retrospective testing of product quality), to a preventive approach to QA, whereby potential hazards are identified and controlled in the pharmaceutical process environment (i.e., prospective ensurance of product quality). HACCP is a logical and cost-effective basis for better decision making with respect to product safety. It provides both manufacturing and quality control (QC) measures with a greater security of control over product safety than is possible with traditional end-product testing.

Beyond this, HACCP can be applied in product establishment (e.g., in the context of outsourcing or in the course of the change of the production place) and in the scale-up of process parameters, and may also support the preparation of product-specific validation plans. Risk analysis will be expanded to the check of documents, including production instructions and test methods, standard operation procedures (SOPs), admittance dossiers, and laboratory reports and specifications. Technical specifications are documents which are needed for the production (filling and packing), check and release of the product, as well as for the specification of qualities of all product components. Additional information which is used to fulfill the requests of different economic circles must also be evaluated. For instance, the United States currently makes more far-reaching demands on product-specific validations than European approval agencies.

In addition, constant quality is required with regard to the use safety of the delivered product and the retention of the specifications for every product. Analysis of the necessary documents and the suggested production process helps to prepare a product-specific validation plan.

THE HACCP CONCEPT

HACCP has been established as a system of food safety assurance based on the prevention of food safety problems, and is accepted by international authorities as the most effective means of controlling foodborne diseases. The HACCP system applied to food safety was developed in the 1960s for the American space programme. It was necessary to design food production processes that ensured the elimination of pathogens and toxins from the foods. As this could not be achieved by finished-product testing alone, the HACCP concept was initiated and has been evolving in the food industry. Since 1973 the U.S. Food and Drug Administration (FDA), followed by the U.S. Department of Agriculture (USDA) and the World Health Organisation (WHO), have encouraged the use of HACCP. Since an effective application of the HACCP concept may help to achieve controlled product safety, systems based on HACCP principles have been incorporated into European Commission (EC) directives on hygiene of foodstuffs and veterinary products.

In the pharmaceutical industry, HACCP can serve as a management tool that provides a more structured and logical approach to the control of identified hazards than that achieved by traditional inspection and QC procedures. Combined with the principles of the FMEA concept, which focuses on engineering systems and processes, it has the potential to identify areas of concern where failure has not yet been experienced. It is therefore useful for the establishment of QA systems. By using HACCP in the pharmaceutical industry, control is transferred from end-product testing alone into the design and control of drug manufacture.

HACCP is a powerful system which can be applied to a wide range of both simple and complex operations in drug manufacture. It is used to ensure product safety at all stages of the production. Implementation of HACCP focuses not only on the product and production methods, but should also apply to the raw material suppliers and final product storage, and must of course consider the general needs of GMP, including qualification, validation and production support operations such as cleaning systems.

The HACCP concept is based on a structured approach which involves the identification and analysis of potential and realistic hazards associated with all stages of pharmaceutical production from raw materials to the distribution of finished products. Microbiological, chemical and physical hazards, documentation, qualification and validation should all be considered if they affect product safety or quality.

Establishment of an Expert Team

Much of the effectiveness of HACCP is achieved through the use of a multidisciplinary team of specialists. The team should have expert knowledge from relevant areas such as quality assurance, microbiology, analytics, production, pharmaceutical technology and engineering.

Product Information

Product-specific data are collected from safety data sheets, application forms, licence agreements, the specification and quality of raw materials and so forth.

Flow Diagram

The manufacturing process is visualised by preparation of a flow diagram. In-process controls are indicated, the equipment used is listed, potential and realistic hazards are analysed and control measures are specified.

Determination of Critical Control Points (CCPs)

CCPs are defined as process steps at which necessary action can be applied to ensure and maintain compliance with specified conditions. CCPs are identified with appropriate measures which can be applied to control each hazard. If no such CCP can be established, product-specific validation should eliminate potential risks of a certain process step. Alternatively, additional subordinated CCPs may be identified or installed.

Definition of Limits

For each CCP, criteria are established which separate acceptability from unacceptability.

Definition of Action Values

Along with defining the limits of CCPs, action values are also established. An action value is a value well below the acceptable limit which allows personnel to react to a potential or realistic hazard prior to the state of unacceptability.

Monitoring

A scheduled sequence of measures, such as in-process controls, observations, testing of product samples and CCP are monitored. With respect to action values and limits, corrective action is applied to ensure specified conditions.

Corrective Action

If the monitoring of CCPs indicates a trend toward loss of control, corrective action has to be defined and initiated.

Exception Report

Uncontrolled conditions, and the need for qualification and/or product-specific validation efforts, identified during HACCP analysis are summarised in an exception report. This report may serve as a product-specific validation plan and as the instrument for a pharmaceutical QA system.

Expert Team

Risk analysis will require the full commitment of management and technical staff to provide the resources necessary for complete analysis of existing and potential risks and subsequent implementation of control measures.

In the HACCP analysis of pharmaceutical processes, a multidisciplinary team of specialists (see Figure 12.1) with expert knowledge of QA, microbiology, analytics, production, pharma-

Figure 12.1. Multidisciplinary team of specialists.

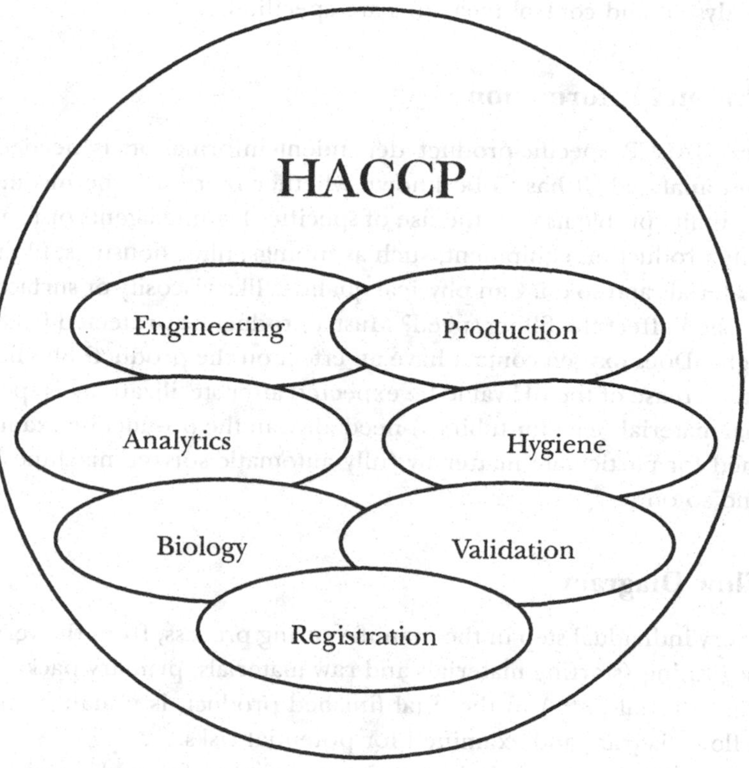

A multi-disciplinary team of specialists with expert knowledge from relevant areas, e.g., quality assurance, microbiology, analytics, production, pharmaceutical technology and engineering, is established for HACCP analysis.

ceutical technology and engineering, product licence, and—of course—issues of GMP will improve the quality of data considered and thus the quality of decisions reached.

The expert team will describe the product and judge specific data that has been collected from safety data sheets, application forms, licence agreements, the specification and quality of raw materials and so on. The chemical structure of active ingredient and raw materials, and the compostion of the product will have to be evaluated. The manufacturing process is visualised by preparation of a flow diagram. In-process controls are indicated, the equipment used is listed, potential and realistic hazards are analysed, and control measures are specified.

Product Information

For HACCP, specific-product dependent information is needed and analyzed. It has to be known whether there may be incompatibility problems with the use of specific cleaning agents or routine production equipment, such as tubings, filter housings, filter materials and so on. Can physical qualities like viscosity or surface tension affect the filling speed? Must a product be protected from light? Does oxygen contact have an effect on the product? Should an increase of the pH value be expected after sterilization? Is special material (e.g., for tubings) needed? Can the product be examined for particulate matter by fully automatic sorting machines? and so on.

Flow Diagram

Every individual step in the manufacturing process, from the very beginning (starting materials and raw materials, primary packaging materials, etc.) to the final finished product, is visualized in a flow diagram and examined for potential risks.

Such a hazard analysis must be undertaken for each product and process and includes, for example, the working areas and personnel situation (hygiene), the flow of materials and the conditions of the surroundings. All these factors combine to give a certain quality of compliance state. The hygiene situation, as the specialized case of a controlled process, can thus be viewed as a closed loop (see Figure 12.2). A given state (hygiene situation) is influenced by a variety of environmental factors (generally speaking, deregulating factors). Regular controls continuously analyse the condition of the hygiene situation—or compliance in general—by comparing nominal and actual states. Corrective measures are enacted as soon as warning values detect a tendency toward divergence from the predefined state. An integrated control system analyses the effects of corrective measures on other

Figure 12.2. Hygiene situation control loop.

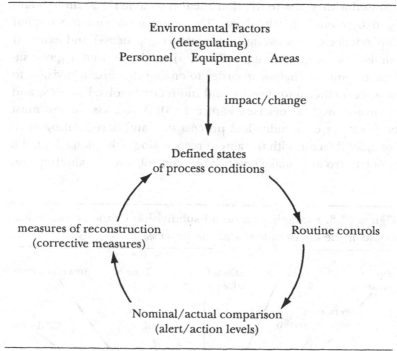

A defined state (hygiene situation) is affected by numerous environmental factors. Routine controls continuously analyse the state of the hygiene situation by comparing the actual and nominal states. Corrective measures come into action as soon as warning values detect a tendency for divergence away from the predefined state.

parts of the system. The appreciation of the importance of microbiological factors in the *manufacture of sterile products in a biologically non-sterile environment* requires a proactive concept and management strategy to ensure constant hygiene conditions.

Determination of Critical Control Points

Critical Control Points (CCPs) are the local conditions, practices, activities, or processes that can be regulated in order to reduce or prevent a known risk. "Control" in this context does not mean checking or testing, but rather the ability to "manage" a potential risk. Each step in a process, for instance a raw material, must be assessed with a view to whether the foreseeable risks can be managed by control mechanisms. The association and the mutual dependence of processing (manufacturing process) and external conditions (environmental influences) on the desired hygiene situation requires analysis in order to enable this broad division to be split further into directly and indirectly involved process and environmental factors (see Figure 12.3). A risk assessment must be made for each individual process step and deregulating environmental factor, with the aim of recognising risks, describing the problem (so as to make it notifiable/controllable), evaluating and

Figure 12.3. Example of a broad subdivision of the connections between the environment and the process.

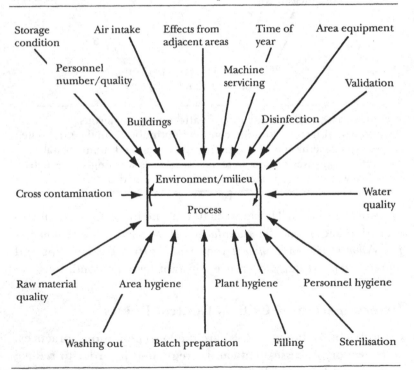

The diagram depicts the connection between and the mutual dependence of process (manufacturing process) and environmental conditions (environmental effects) on the prevailing hygiene situation. Direct and indirectly involved process and environmental factors are illustrated.

eliminating it or—if this is impossible—constructing suitable management mechanisms.

Via the formalized approach of the HACCP concept, each individual step or component in the process is broken down into product-specific or general (e.g., plant-specific) risks, particularly but not only those relating to hygiene. Such a procedure also enables the magnitude of potential risks to be estimated, warning or threshold values for control tests to be defined and suitability criteria to be specified.

Risks that may not necessarily lead to hygiene risks in the strictest sense must be incorporated and managed. Their importance is assigned a relative weight whereby, for instance, account is taken of the interests of various specialist departments. The departments affected are jointly responsible for this evaluation process. If risks cannot be fully excluded, then appropriate management strategies such as product-specific validation must be developed and established. The aim must always be the safety and compliance of the entire manufacturing process.

Two types of CCPs should be distinguished: As an *absolute* effective control point, thermal treatment or pH reduction excludes a risk in respect of hygiene terms, whereas cooling cannot be regarded as suitable for complete control of a risk. Cooling, therefore, is one case where a risk can be *reduced* (e.g., prevention of growth of microorganisms), but not completely prevented.

Product-specific validation of certain process steps may have to be established at CCPs at which monitoring is not feasible or insufficient, in order to control a potential or realistic hazard.

Definition of Limits and Action Values

Limits serve as criteria with which to determine whether a process is under control at a given control point. To enable early and stepwise action to be taken when unusual events occur, warning values and alarm/emergency plans must be defined. The effectiveness of such established measures is to be investigated and validated. The limits for physical, chemical or (micro)biological parameters, for example, are to be given (temperature, time, pH) and their prognostic relevance evaluated.

Monitoring Program and In-Process Control

The statistical analysis of microbiological and particulate monitoring over long periods and during both production and non-operating conditions of areas, personnel, machinery and equipment, production surroundings and process water leads to an accurate insight into the hygiene status of the production factors. Observation of trends enables the early adoption of countermeasures. A preventive, stepwise reaction to changes is thereby possible, with risk reduction and risk management always preferable to damage limitation.

In-process controls include the observation of the listed critical limits, and are to be regarded as warning processes which allow correction of a process that has run out of control, even during the course of that process (e.g., batch analysis, dry run of a filling line after a long idle period, print and code colour quality). Here too, limits and prognostic significance must be assessed.

The undertaking of in-process controls is thus the decisive operation and the core of the HACCP concept. An analysis should be carried out to determine when corrections are to be made and who is competent to make or initiate them.

Establishment of Corrective Action

Observation of the specified critical limits is to be considered as a warning procedure. Even during an ongoing process, corrective measures are to be initiated as soon as the in-process controls indicate that the process at a CCP is getting out of control. Such measures are to be laid down and their capacity to return the process to a permissible status within a specified time should be evaluated.

The designed CCP system must be functioning correctly. For proof, a system check should be introduced, for example by investigating random samples of end product or analyses at certain CCPs (e.g., bioburden analysis) within validation batches. Checking the process environment (microbiological monitoring, in-process control) is also a part of the confirmatory system.

Exception Report

For each individual production step, risk control mechanisms have been defined which aid in checking the analysed risk. If a manufacturing step classified as critical turns out not to be in a controlled status, product-specific measures should be defined and taken in order to reach a controlled state of a potential risk. The exception report may then serve as product-specific validation plan.

EXAMPLE OF HACCP ANALYSIS

The HACCP concept is logically applicable to all steps of the manufacturing process and pharmaceutical production system in general, including checks on purchased goods and materials, weighings, batch preparation, sterile filtration, filling, autoclaving, packaging, storage and dispatch. From the production flow diagram (Figure 12.4) of the manufacture of a parenteral terminally sterilized drug, an example of HACCP analysis is depicted (in Table 12.1).

Each step in the process is subdivided into its individual measures (column 1) and analyzed with respect to the risks to be considered (column 2). Column 3 shows the parameter by which a risk can be managed. The risk is evaluated in column 4, taking into account whether a risk is controlled or not. A risk can be controlled through organizational measures (SOPs, etc.), validation studies, or through associated control points. The last column lists relevant control methods and documentation, such as validation studies, SOPs and process instructions.

Figure 12.4. Flowchart of production of a parenteral drug.

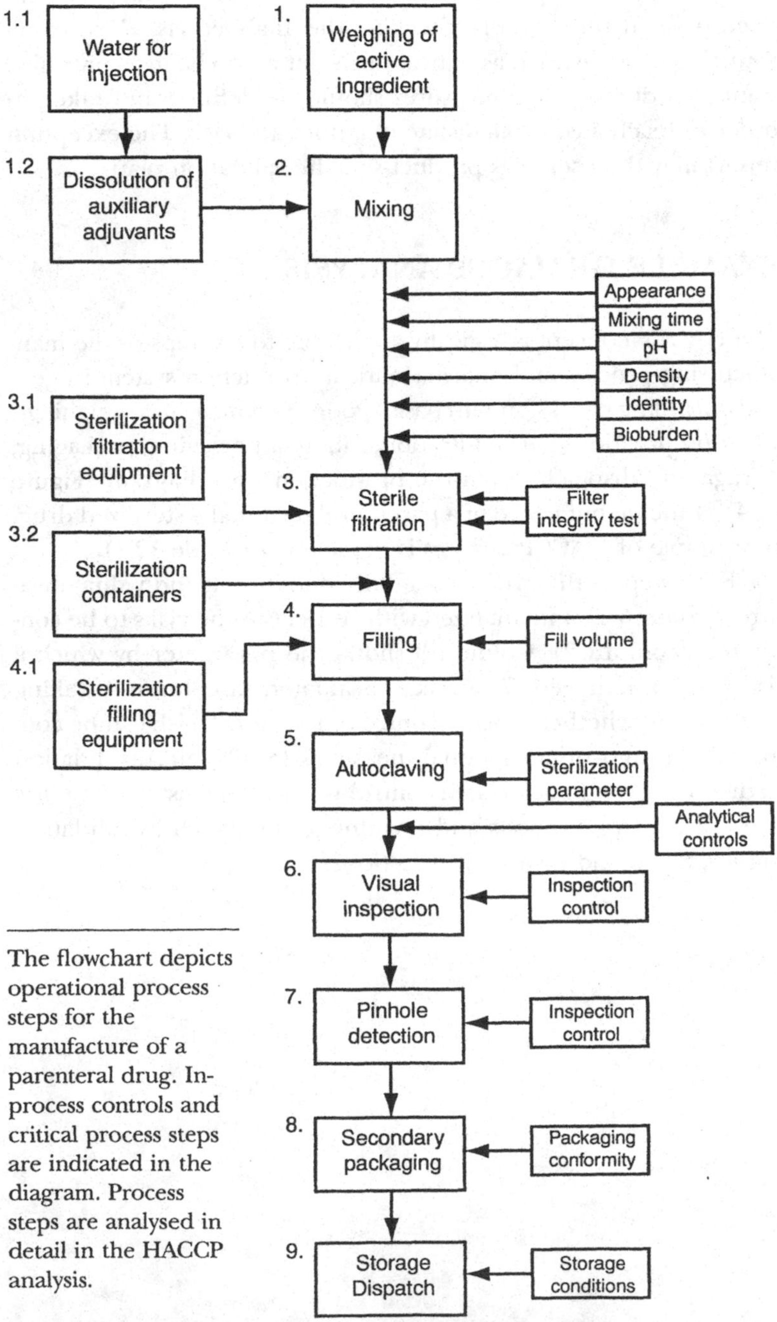

The flowchart depicts operational process steps for the manufacture of a parenteral drug. In-process controls and critical process steps are indicated in the diagram. Process steps are analysed in detail in the HACCP analysis.

Table 12.1. HACCP analysis of a parenteral terminally sterilized drug.

Production step Single measures	Risk analysis	Critical Control Points	Evaluation	Documents
Starting material				
Active ingredient	Reproducible quality	Vendor audit	critical uncontrolled	PVP
Auxiliary adjuvant	Reproducible quality	Vendor audit/ ISO Certification	critical controlled	SOP USP 23
Product container	Reproducible quality	Vendor audit	critical controlled	SOP
Primary packaging Glass				
Chemical-physical features	Release of ions and impurities	Goods incoming for inspection, glass of internal surface treatment	critical uncontrolled	PVP
	Release of Na$_2$O in product solution	Goods incoming for inspection	critical uncontrolled	SOP
	Breakability	Goods incoming for inspection	critical uncontrolled	SOP
		Optical control of the finished product	critical controlled	SOP
		Leakage testing on the finished product	critical controlled	SOP

Continued on the next page.

Continued from the previous page.

Production step Single measures	Risk analysis	Critical Control Points	Evaluation	Documents
	Light protection	Goods incoming for inspection	critical uncontrolled	PVP
Impurity	Particulate risks	Qualification of the washing machine	critical controlled	Validation
	Biological risks	Qualification of sterile/ depyrgenious tunnel	critical controlled	Validation
	Particulate matter during shelf life	Stability data	critical uncontrolled	PVP
Pretreatment	Change of features of the primary packaging (inner surface treatment)	Goods incoming for inspection, evaluation of product-specific impacts	critical uncontrolled	PVP
Defects	Breakage of glass during process	Goods incoming for inspection	critical controlled	SOP
Geometry	Limited processability	Goods incoming for inspection	critical controlled	SOP
Ring coding	Mixup during process prior to labelling	Color indicates product	critical controlled	Batch record

Continued on the next page.

Continued from the previous page.

Production step Single measures	Risk analysis	Critical Control Points	Evaluation	Documents
Weighing				
Preparation	General cleanliness within the warehouse	Cleaning documentation	critical controlled	SOP
	Mixup of containers	Procedual regulation	critical controlled	SOP
Procedure	Cleaning of the containers	Cleaning documentation	critical controlled	SOP
	Mixup of starting materials	Control by second person	critical controlled	SOP
	Malfunction of scales	Hierarchical levels: Office of Weights and Measures, maintenance, daily calibration	critical controlled	SOP
	Multiple weighing of starting materials	Setup analysis	critical controlled	TP Batch record
	Omission of starting material	Setup analysis	critical controlled	TP Batch record
	Application of different starting material in one batch	Batch-specific weighing	critical controlled	SOP

Continued on the next page.

Continued from the previous page.

Production step Single measures	Risk analysis	Critical Control Points	Evaluation	Documents
Balancing	Weight control	Control of conformity of release number/ batch and factor	critical controlled	SOP Batch record
	Risk of contamination:			
	a) machinery/equipment cleaning procedure	Procedual regulation: batch-specific weighing and cleaning of equipment	critical controlled	SOP Batch record
Contamination control	b) personnel training	Procedual regulation: training/hygienic training	critical controlled	SOP
	access	Access for authorized personnel only	critical controlled	SOP
	c) room			
	room class B (10.000)	Regular control	critical controlled	SOP
	hygienic condition	Regular control	critical controlled	SOP
	d) impurity through extraneous substances	Procedual regulation: protection against mix-up	critical controlled	SOP
	e) micro-organisms in	Bioburden: test on pathogenic micro-organisms		
	active and auxiliary	Specifcation of maximum bioburden	critical controlled	SOP
	adjuvants			Batch record

Continued on the next page.

Continued from the previous page.

Production step Single measures	Risk analysis	Critical Control Points	Evaluation	Documents
Set up				
Batch preparation	Active ingredient not completely soluted	Temperature control	critical controlled	Batch record
		Subordinated control point: batch analysis on active ingredient	critical controlled	TP
	Incomplete homogenisation of active ingredient	Subordinated control point: batch analysis on active ingredient	critical controlled	TP
Batch size	Setup procedure not reproducible	Determination of batch size and vessel	critical controlled	Batch record
Starting material	Batch-extraneous starting material	Control and registration within batch documentation	critical controlled	SOP
		Subordinated control point: set-up analysis	critical controlled	SOP
Setup vessel	Cross-contamination	Documentation of the intermediate product; cleaning evidence	critical controlled	SOP Batch record

Continued on the next page.

Continued from the previous page.

Production step Single measures	Risk analysis	Critical Control Points	Evaluation	Documents
Process water	High bioburden	Inspection of the water source (WfI)	critical controlled	SOP
Oxygen content	Nitrogen impact insufficient	In-process control: oxygen content	critical controlled	SOP Batch record
	Calibration of equipment irregular	Calibration procedure	critical controlled	SOP
Homogenization	Homogenization insufficient	In-process control: visual condition	critical controlled	SOP
	Homogenization not reproducible	Product specific validation: setup procedure (mix time)	critical controlled	PVP
Content	Concentration of active ingedient irregular	Setup analysis	critical controlled	TP
	Minor content auxiliary adjuvant	Setup analysis	critical controlled	TP
Bioburden	High bioburden in bulk solution	Determination of bioburden	critical controlled	SOP
Holding time	Growth promotion due to product	Product-specific validation, growth promotion/conservation testing procedure	critical uncontrolled	PVP

Continued on the next page.

Continued from the previous page.

Production step Single measures	Risk analysis	Critical Control Points	Evaluation	Documents
Contamination	Risks of contamination			
	a) equipment			
	cleaning procedure	procedural regulation	critical controlled	SOP Batch record
	b) personnel			
	training	Training/hygenic training	critical controlled	SOP
	access	Access for authorized personnel only	critical controlled	SOP
	hygiene	Follow-up bioburden analysis after germ reduction filtration	critical controlled	SOP
	c) room			
	room class B (10.000)	Regular monitoring	critical controlled	SOP Batch record
	cleaning procedure	Cleaning documentation	critical controlled	Validation Batch record
	hygenic condition	Regular control	critical controlled	SOP Batch record

Continued on the next page.

Continued from the previous page.

Production step Single measures	Risk analysis	Critical Control Points	Evaluation	Documents
	d) impurity through extraneous substances			
	cross-contamination	Procedural regulation: protection against mixup	critical controlled	SOP
	chemical impurity during sampling	Procedural regulation: sampling and hygiene	critical controlled	SOP
	microbial impurity upon sampling	Procedural regulation: sampling and hygiene	critical controlled	SOP
Filtration				
Particles	Decreased product quality	Final inspection: optical control	critical controlled	SOP Batch record
	Filter size/filter surface unsuitable	Determined batch size	critical controlled	Batch record
Reduction of micro-organisms	Filter material unsuitable/ pore size influenced by ingredients	Product-specific validation	critical controlled	PVP

Continued on the next page.

Continued from the previous page.

Production step Single measures	Risk analysis	Critical Control Points	Evaluation	Documents
	Bacterial retention rate inadequate	Product-specific validation	critical uncontrolled	PVP
	Reproducibility of filter material uncertain	Vendor audit/Validation	critical controlled	SOP
	Bioburden too high	Tolerance before/after filtration	critical controlled	SOP
	Filtration pressure too high	Self-regulating system	critical controlled	Validation
Filter integrity	Filter defect	Integration testing before/after utility	critical controlled	SOP
		Determination of min/max filtration time	critical uncontrolled	PVP
	Filter unsterile	Validation of equipment sterilisation	critical controlled	Validation
	Filter blocked	Change of filter/further filtration (procedual regulation)	critical controlled	SOP
	Forerun minor content	Procedual regulation: forerun = 10–15 times exchange of filter and tube volume	critical uncontrolled	PVP

Continued on the next page.

Continued from the previous page.

Production step / Single measures	Risk analysis	Critical Control Points	Evaluation	Documents
Holding time	Adsorption, extraction	Product specific validation	critical uncontrolled	PVP
	Growth of microorganisms or loss of active ingredient during holding time	Validation of holding time	critical uncontrolled	PVP
Contamination		Subordinated control point: sterilisation of filled containers at 121°C/15 min.	critical controlled	Batch record
	risk of contamination			
	a) equipment			
	cleaning and sterilisation	Procedual regulation	critical controlled	SOP
	b) personnel			
	training	Procedual regulation: training/hygenic training	critical controlled	SOP
	hygiene	Aseptical operation, procedural regulation	critical controlled	SOP
	c) room			
	room class B (10.000)	Control/process validation	critical uncontrolled	PVP
	hygenic condition	Control/process validation	critical uncontrolled	PVP

Continued on the next page.

Continued from the previous page.

Production step Single measures	Risk analysis	Critical Control Points	Evaluation	Documents
	d) impurity through extraneous substances cross-contamination	Procedual regulation: protection against mixup Follow-up inspection: finished product analysis	critical controlled	SOP
	e) pressure filtration nitrogen microbial contamination	Regular sampling of Inertgas system Follow-up sterile filtration 0.2 μm	critical controlled	SOP
Filling Ring code	Mixup of products	Documentation within manufacturing protocol/IPC	critical controlled	SOP
Cleaning procedure	Incomplete cleaning of ampoule (chemical-physical impurity)	Water pressure	critical controlled	Validation
		Compressed air	critical controlled	Validation
		Applicability of spray needles	critical controlled	Validation
	Microorganisms/impurity through rinsing water	Regulation upon negative ICP-results	critical controlled	SOP
		Process water: water for injection		Batch record

Continued on the next page.

Continued from the previous page.

Production step / Single measures	Risk analysis	Critical Control Points	Evaluation	Documents
Depyrogenisation	Endotoxins present	Validation of depyrogenation tunnel	critical controlled	Validation
Filling	Growth of microorganisms through filling line	Qualification of the filling line (media fill)	critical controlled	Validation
Forerun	Contaminated ampoules	Qualification sterile tunnel	critical controlled	Validation
	Minor content of active ingredient	Process validation	critical uncontrolled	PVP
Dosage	Difference in filling Batch record	Generally to be controlled according to GMP	critical controlled	SOP
	Difference in extractable volume	Control of filling volume via in-process control	critical controlled	SOP
		Finished-product testing analysis/optical control	critical controlled	Batch record
Filling needle	Diameter of filling needle unsuitable	In-process control	critical controlled	SOP
Sealing off	Cracks, blowed spikes, sealing off length	Adjustment of the filling line/ In-process control	critical controlled	SOP

Continued on the next page.

Continued from the previous page.

Production step Single measures	Risk analysis	Critical Control Points	Evaluation	Documents
	Leakage	Density testing	critical controlled	SOP
	Impurity through the process	In-process control: burner adjustment, visual inspection, avoidance of particles	critical controlled	SOP
		Optical control of finished product	critical controlled	SOP
Contamination		Risks of contamination		
	a) equipment			
	cleaning and sterilisation	Qualification of product vessel	critical controlled	Validation
		Qualification of the aeration system	critical controlled	Validation
		Qualification of filling line (growth media filling)	critical controlled	Validation
		Qualification sterilisation of production equipment	critical controlled	Validation

Continued on the next page.

Continued from the previous page.

Production step Single measures	Risk analysis	Critical Control Points	Evaluation	Documents
	b) personnel			
	training	Procedual regulation: training/hygenic training	critical controlled	SOP
	hygiene	Production-specific control during validation	critical uncontrolled	PVP
	c) room			
	room class B (10.000)	Production specific control during validation	critical uncontrolled	PVP
	hygienic condition	Production-specific control during validation	critical uncontrolled	PVP
	filling in room class A (100)	Production-specific control during validation security safety laminar flow system	critical uncontrolled	PVP
	d) impurity through extraneous substances cross-contamination	Cleaning validation	critical controlled	Validation
		Follow-up control: finished product analytics	critical controlled	SOP Batch record

Continued on the next page.

Continued from the previous page.

Production step Single measures	Risk analysis	Critical Control Points	Evaluation	Documents
Sterilisation				
Transport	Physical damage of containers	Optical control	critical controlled	SOP
Storage	Mixup within different products	Ring code/organized procedure	critical controlled	SOP
	within sterile/non-sterile goods	Work-in-process indicator cards/tray labels, sterilisation indicators	critical controlled	SOP
	same products of different batches	Ch.B./work-in-process indicator cards/tray labels, registration of the autoclave loads	critical controlled	SOP
Holding time	Growth of microorganisms	Process time/product-specific processing time	critical uncontrolled	PVP
Loading	Mixup	batch-specific sterilisation Line clearance procedures	critical controlled	SOP
	Subnormal temperature	Check of position of product thermo couples	critical controlled	SOP

Continued on the next page.

Continued from the previous page.

Production step / Single measures	Risk analysis	Critical Control Points	Evaluation	Documents
	Deregulated product detector	integrity testing during and after the sterilisation cycle	critical controlled	SOP
	Security of sterilisation	worst case qualification of the autoclave	critical controlled	Validation
Temperature penetration	Product influences temperature penetration respectively, distribution	detection of the comparability of the worst case with the product	critical uncontrolled	PVP
Temperature	Full-load/partial-load production batches influence temperature distribution/penetration	validation using full-load/partial-load chamber	critical uncontrolled	PVP
	Wrong sterilisation cycle	Set/actual value comparison within the batch record and shift leaders	critical controlled	Batch record SOP
Cooling	Non-sterile air	Air filter validation	critical controlled	Validation
	Microorganisms in cooling water	Microbiological monitoring	critical controlled	SOP
		Subordinated control point: integrity testing of product containers	critical controlled	SOP

Continued on the next page.

Continued from the previous page.

Production step Single measures	Risk analysis	Critical Control Points	Evaluation	Documents
	Leakage	Subordinated control point: integrity testing of product containers	critical controlled	SOP
Transport	Physical damage	Optical control	critical controlled	SOP
	Mixup			
	within different products	Ring coding/organized process	critical controlled	SOP
	the same products of different batches	Ch.B./work-in-process identification cards/tray labels	critical controlled	SOP
		registration of the autoclave loads		
		ring code identification during labeling		
Final testing				
Particles	Visible particles	Particle control	critical controlled	TP
Content conformity	Over/under dosage	In-process control, finished product control	critical controlled	TP Batch record
	Over/under content of active ingredient	Finished product control	critical controlled	TP SOP
	Byproducts/degradation products	Finished product control	critical controlled	TP SOP

Continued on the next page.

Continued from the previous page.

Production step Single measures	Risk analysis	Critical Control Points	Evaluation	Documents
Method validation (analytics)	Reproducibility of the analysis	Product-specific method validation	critical uncontrolled	PVP
Sterility	Non-sterility	Product-specific method validation	critical uncontrolled	PVP
		Validation autoclave/finished-product control	critical controlled	Validation
		Incubation 14 days, 2 temperatures	critical controlled	SOP
Optical control				
Feeding	Mixup	Batch-specific feeding Line clearance	critical controlled	SOP
Particles	Unsuitable machine adjustment	Validation of the equipment Set/actual value comparison within batch record	critical controlled	Validation
	Insensitive detection of quality defects	Product-specific validation	critical uncontrolled	PVP
Leakage testing	Inadequate testing procedure (conductivity)	Product-specific evaluation	critical controlled	SOP

Continued on the next page

Continued from the previous page.

Production step Single measures	Risk analysis	Critical Control Points	Evaluation	Documents
	Incomplete registration of leakage	Validation of equipment	critical controlled	Validation
Labels		Product-specific validation	critical uncontrolled	PVP
Bulk labels	Variable data incorrect/ uncontrolled	Material number/administration of version number Release procedure	critical controlled	SOP Batch record
Container labels	Variable data incorrect/ uncontrolled	Material number/administration of version number Release procedure	critical controlled	SOP Batch record
Carton labels/ additional labels	Unauthorized printing	Documentation within list of material Documentation in batch record Unique regulation of printing procedure	critical controlled	SOP
Palette container labels	Unauthorized printing	Documentation within list of material Documentation in batch record Unique regulation of printing procedure	critical controlled	SOP Batch record

Continued on the next page.

Continued from the previous page.

Production step Single measures	Risk analysis	Critical Control Points	Evaluation	Documents
Storage				
Storage area	Mixup of products	Batch-specific storage	critical controlled	SOP
Access	Unauthorized disposition	Access limitation	critical controlled	SOP
Intermediate storage	Mixup (batches/ extraneous substances)	Labeling of outer cartons/ palettes, organized/procedual regulation	critical controlled	SOP
Dispatch				
Labeling	Wrong labels	In-process control, batch-specific Ch.B.	critical controlled	SOP
Packaging	Breakage	Packaging in cartons/ palette wrapping	critical controlled	SOP
Recipient	Identification	Carton/palette label: batch number, material number	critical controlled	SOP
Temperature control	Precipitation, breakage, instability	Temperature control and adjustment within storage area and on transport	critical uncontrolled	PVP

Continued on the next page.

Continued from the previous page.

Production step Single measures	Risk analysis	Critical Control Points	Evaluation	Documents
Reconciliation				
Active ingredient	No batch-specific use of active ingredient/reconciliation upon complaints	Procedural regulation/ reconcil- batch record	critical controlled	SOP Batch record
Auxiliary adjuvant	No batch-specific use of ingredients/reconciliation upon complaints	Reconciliation through electronic data processing	critical controlled	SOP
Primary packaging	No batch-specific use/ reconciliation upon complaints	Reconciliation through electronic data processing	critical uncontrolled	SOP
Bulk labels	No batch-specific use/ unauthorized use	Procedural regulation/ batch record	critical uncontrolled	SOP Batch record
Filled containers	Reconciliation upon complaints	Reconciliation through electronic data processing	critical uncontrolled	SOP

SOP: Standard Operating Procedure; relevant documents are listed in this column.

TP: Test Procedure; relevant methods are listed in this column.

Validation: Validation protocol documents and validation report documents are listed in this column.

Batch report: Reference to the batch manufacturing protocol; relevant data, e.g., on in-process control, are introduced into this document.

PVP: Product Validation Plan; uncontrolled issues of HACCP analysis must be referred to in a product-specific validation plan.

PRODUCT VALIDATION PLAN (PVP)

For each individual production step, risk control mechanisms have been defined which serve to check the analysed risk. If a manufacturing step classified as critical turns out not to be in a controlled status, measures must be taken in order to reach a controlled condition. In the example of HACCP, product-specific measures have been defined which are to be taken to reach a controlled state of a potential risk.

Inspection/Self-Inspection of the Synthesis of the Therapeutically Active Substance

In order to guarantee and evaluate the reproducible quality of the used therapeutically active starting material in the manufacturing process, vendor audit/self-inspection should be carried out according to cGMP requirements.

Inspection of Sub-Suppliers

In order to guarantee and evaluate the reproducible quality of primary and secondary packaging materials, and auxiliary materials used in the manufacturing process, sub-suppliers should be judged.

Stability Data

A report dealing with the product quality affected by the kind of glass used as primary packaging material should be worked out. Stability test data and information derived from product development should be evaluated. The interaction of product and primary packaging material (e.g., proof of contaminations) can be analysed by such examination or evaluation.

Light Protection

The influence of light on the chemical stability of a product may have to be analysed, if light protection is indicated by chemical properties of the active pharmaceutical ingredient.

Bulk Preparation/Homogenization

The homogenization time is to be validated product-specifically.

Growth Promotion Test (Product-Specific Holding Times)

In view of holding times during processing, growth promotion properties of the product on microorganisms should be evaluated. Product-specific evaluation (e.g., tests on preservative efficiency) should reflect on standard reference cultures an *in-house* population.

Bioburden of Bulk Preparation and Holding Time

Within the testing of starting materials and bulk preparation, the microorganisms present should be identified. This enables a comparison between product-specific in-house population and method validation using reference strains proposed by the respective pharmacopoeia.

Filtration (Bacterial Retaining Rate)

The bacterial retaining rate of the filter material should be validated product-specifically. The process parameters of filtration (pressure, batch size, process time and so on) will have to be reflected in examination of bacterial retention rate.

Filtration (Extraction Behaviour)

The extraction behaviour of sterile filters while they are in contact with the product should be evaluated.

Filtration (Specification of Filtration Time)

The filtration time may, at standard batch size, indicate blockages or reduced filtration rates. As criterion for an in-process control, the filtration time of the product should be analysed on the maximum batch size. On the basis of this analysis, a minimum and maximum filtration time should be specified in the batch manufacturing record.

Residual Humidity in the Manufacturing Equipment

If the manufacturing equipment is sterilised a residual humidity can remain, which may have an influence on the content of the active ingredient. This is taken into account by rejecting filled product containers from the pre-run. In order to facilitate a product-specific statement on the problem of a potential undercontent of active ingredient, the first containers filled should be analysed for the content of the active ingredient.

Particle Situation in Class A/100 and Class B/10.000

The state of the particle situation in class A/100 and B/10.000 areas should be checked within the validation of the manufacturing process or established as routine environmental control.

Hygiene Situation in Class A/100 and class B/10.000 Areas

The hygiene situation in the area of class A/100 and B/10.000 should be checked within the validation of the manufacturing process or established as routine environmental control.

Sterilization

The product sterilization parameters should be validated in view of temperature distribution and temperature penetration. The heat distribution is recorded in a full chamber load of the autoclave. If indicated, product-specific validation of partial chamber loads of the autoclaves shuld be performed.

Mixup of the Product

Since a product may be unlabelled during the manufacturing process, measures have to be established in order to identify the product during work-in-process conditions such as internal transport, conduction of optical control, leak testing and storage.

Method Validation

The reproducibility of both chemical and biological methods applied to the product should be evaluated. The method for determining the total viable count in bulk preparations should be validated product-specifically, using standard reference cultures and in-house population.

Optical Inspection

The quality of the optical inspection of filled product containers should be validated product-specifically.

Leak Testing

The machine check for the absence of leaks in product containers on the basis of conductivity test should be validated product-specifically (adjusting sensitivity/safety evaluation).

Temperature Mapping while Storing and Dispatching Products

In order to document and evaluate the conditions while the product is stored and transported, temperature sensors should be introduced into the product load.

SUMMARY

The HACCP concept implements a tool for checking and evaluating the pharmaceutical manufacturing processes and any regulated process steps in terms of potential or actual risks affecting the product. Before suitable process controls are established, the manufacturing process has to be analysed in terms of its critical steps, which must then be carefully monitored. In the pharmaceutical industry, HACCP focuses on biological (e.g., hygienic situation, chemical, technical and regulative) impact on the total pharmaceutical process. The formalised procedure discussed in this review will aid in an effective strategy for risk analysis, the establishment of monitoring systems and the evaluation of product-specific validation efforts.

Hazard analysis and the monitoring of Critical Control Points identified in such an assessment lead to the establishment of an effective concept for QA.

REFERENCES

Amtsblatt der Europäischen Gemeinschaft. 1993. Richtlinie 93/43/EWG des Rates vom 14. Juni 1993 über Lebensmittelhygiene. *Nr.L* 175:1–11.

Anonymous. 1998. Current Good Manufacturing Practice in manufacturing, processing, packing or holding of drugs/Current Good Manufacturing Practice for finished products. Code of Federal Regulations 21 CFR, Parts 210/211.

Anonymous. 1991. The principles and guidelines of Good Manufacturing Practice for medicinal products for human use. Commission Directive 91/356/EEC.

Anonymous. 1998. *Medical devices—Risk analysis.* EN 1441 1998. Berlin: Beuth Verlag.

Anonymous. 1994. Special issues on HACCP: Basic principles, application and training. *Food Control* 5 (3):131–209.

Betken, R. 1997. Eine Symbiose von ISO und HACCP. *Zeitschrift für Lebensmitteltechnik* 48 (9):24–29.

Bogatz, C., and J. Merdian. 1997. HACCP und Qualitätsmanagement—perfekte Systemsynergien. *Journal for the Food Industry* 48 (5):30–33.

Buntain, B. 1997. The role of the food animal veterinarian in the HACCP era. *Journal of the American Veterinarian Medical Association* 210 (4):492–495.

Cross, H. R. 1996. International meat and poultry HACCP alliance. Journal of the American Veterinarian Medical Association 209 (12):2048.

Cullor, J. S. 1996. An HACCP learning module for graduate veterinarians. *Journal of the American Veterinarian Medical Association* 209 (12):2049–2050.

EG-Leitfaden einer Guten Herstellungspraxis für Arzneimittel, 4th ed. 1995. Ed. G. Auterhoff. Aulendorf: Editio Cantor Verlag.

International Commission on Microbiological Specifications for Foods (ICMSF). 1988. Application of the hazard analysis critical control point (HACCP) system to ensure microbiological safety and quality. *Microorganisms in Foods* 4. Oxford: Blackwell Scientific.

Isoard, P., D. Vidal, F. Thibault, and G. Ducel. 1997. Biodecontamination, normalisation europeenne et industrie pharmaceutique. *Annales Pharmaceutiques Francaises* 51 (4):186–196.

Jahnke, M. 1997. Use of the HACCP concept for the risk analysis of pharmaceutical manufacturing processes. *European Journal of Parenteral Sciences* 2 (4):113–117.

Jahnke, M. 1998. Bacillus cereus PHG 5/11 in the microbiological process control of sterilisation methods. *European Journal of Parenteral Sciences* 3 (1):5–8.

Kohlert, E. U., and A. Kohrt. 1998. Pharmaceutical quality assurance, production and quality control as columns of a quality management system. *Pharmaceutische Industrie* 60 (3):200–204.

Kieffer, R. G., S. Bureau, and A. Borgmann. 1997. Applications of Failure Mode Effect Analysis in the pharmaceutical industry. *Pharmaceutical Technology Europe* 9:36–49.

Leaper, S., ed. 1992. *HACCP: A practical guide*. Technical Manual No. 38. Campden Food and Drink Research Association.

Ljungqvist, B., and B. Reinmüller. 1995. Hazard analyses of airborne contamination in clean rooms—Application of a method for limitation of risks. *PDA Journal of Pharmaceutical Science and Technology* 49 (5):239–243.

Ljungqvist, B., and B. Reinmüller. 1995. Qualification of weighing station for pharmaceutical substances. *Journal of Pharmaceutical Science and Technology* 49 (2):93–98.

Mayes, T. 1992. Simple users' guide to the hazard analysis control points concept for the control of food microbiological safety. *Food Control* 3 (1):14–19.

McDermott, R. E. 1996. The basics of FMEA. *Quality Resources.* New York.

Mitchell, R. T. 1992. How to HACCP. *British Food Journal* 94 (1):16–20.

Mortimore, S., and C. Wallace. 1994. *HACCP: A practical approach.* London: Chapmann and Hall.

Pierson, M. D., ed. 1993. HACCP. *Grundlagen der produkt- und prozessspezifischen Risikoanalyse.* Hamburg: Behr's Verlag.

Reed, C. A. 1995. FSIS explains benefits of HACCP (news). *Journal of the American Veterinarian Medical Association* 207 (3):284.

Royal Institute for Public Health and Hygiene. 1995. *HACCP Training Standard: HACCP principles and their application in food safety.* London: RIPHH.

Sahni, A. 1993. Using Failure Mode and Effect Analysis to improve manufacturing processes. *Medical Device and Diagnostics Industry* 47–51.

Schöne, D. 1998. Lagerung und Transport kühlpflichtiger Arzneimittel. *Deutsche Apotheker Zeitung* 138 (12):53–58.

Seyfarth, H. 1997. Mikrobiologische Gesichtspunkte bei der Arzneimittelherstellung. *Swiss Pharma* 19 (11):17–35.

Sperber, W. H. 1991. Use of the HACCP system to assure food safety. *Journal—Association of Official Analytical Chemists* 74 (2):433–440.

Stamatis, D. H. 1995. *Failure Mode and Effect Analysis.* Milwaukee, WI: ASQC Quality Press.

World Health Organisation. 1995. Hazard analysis critical control point system, concept and application. Report of a WHO Consultation with the participation of FAO 29–31 May 1995. WHO/FNU/FOS/95.7.

World Health Organisation. 1995. Training aspects of the hazard analysis critical control point system (HACCP). Report of a WHO workshop on training in HACCP with the participation of FAO. Geneva, 1–2 June 1995. WHO/FNU/FOS/96.3.

Index

Printed in the United States
by Baker & Taylor Publisher Services